* * *

Essentials of

Anatomy & Physiology

A Review Guide for Therapists, Yoga Teachers, Holistic Health Practitioners, Healers & Wellbeing Coaches

Module 2

The Muscular System, Parts I & II

The Cardiovascular System, Parts I & II

Amanda Jackson-Russell, Ph.D.

* * *

* * *

Copyright © Amanda Jackson-Russell 2020.

The contents of this book (including the text and all illustrations and diagrams, and the cover design, illustrations and text) are the copyright of the author Amanda Jackson-Russell, in accordance with the Copyright, Designs and Patents Act 1988.

All rights reserved. No part of this publication may be reproduced, stored in a retrieval system, or transmitted in any form or by any means, electronic, mechanical, photocopying, recording or otherwise without the prior written permission of the author, except in accordance with permissions stipulated in the Copyright, Designs and Patents Act 1988 (such as those related to use of extracts by educational establishments).

First published January 2020.
First edition January 2020.

* * *

Contents

Introduction to the Book Series	iv
Module 2	1
Section 1: The Muscular System – Part I	3
Section 2: The Muscular System – Part II	13
Section 3: The Cardiovascular System – Part I	31
Section 4: The Cardiovascular System – Part II	41
Answers to Quiz Questions – Module 2	55
About the Author	56
Other Modules in this Series	57

Introduction to the Book Series

This is the First Edition of the book series *Essentials of Anatomy & Physiology A Review Guide for Therapists, Yoga Teachers, Holistic Health Practitioners, Healers & Wellbeing Coaches* by Amanda Jackson-Russell, Ph.D.

This book series is designed to provide essential information on the structures (anatomy) and functions (physiology) of the human body (commonly referred to as A&P) in an easy-to-read and review* format for qualified therapists, yoga teachers, holistic practitioners, healers, wellbeing coaches, and teachers, students and prospective students of related professional training. It is not designed to be a comprehensive textbook on the subject, as there are already many good texts available in differing levels of depth.

The series is aimed at four main audiences: 1) qualified practitioners who have already undertaken training in A&P, but subsequently find that certain essential information has "slipped their mind" and they require a quick refresher; 2) prospective student practitioners and teachers who are considering undertaking a professional qualification that will require knowledge of A&P, to give them an easy-to-read and "non-threatening" introduction (and to help them avoid potentially getting "bogged down" in complex details); 3) student practitioners and teachers, as an exam review* aid; and 4) teachers of relevant professional or introductory courses, for use as an information review* and teaching aid. The series may also be useful in several other contexts, such as being a quick reference in clinic or teaching situations.

The book series is published as a series of modules that cover all of the main anatomical structures and physiological systems of the human body. Each module covers two or more of these main areas, depending on the complexity of the particular areas and depth of information deemed appropriate. The entire book series thus covers: Cells and Tissues, The Skin, The Skeletal System, The Muscular System, The Cardiovascular System, The Respiratory System, The Nervous System, The Digestive and Eliminative System, The Reproductive System, The Urinary System, The Endocrine System, The Lymphatic System and The Immune System.

For ease of quick reference, the educational information is presented, as far as possible, in a highly visual format. In each topic section, information is sometimes repeated in different ways and is also presented towards the end of the section as a summary. Finally, each section ends with a list of quiz questions. The answers to quiz questions are given at the end of the particular module (and ideally should not be read until after the reader has genuinely attempted to answer the quiz questions!).

*The word "review" is used here in the American sense, as in "reviewing course material for an exam". In the United Kingdom, the more commonly used words would be "revision" or "revise", as in "revising for an exam" or "doing revision for an exam".

Module 2

1. The Muscular System – Part I

2. The Muscular System – Part II

3. The Cardiovascular System – Part I

4. The Cardiovascular System – Part II

* * *

Section 1

The Muscular System – Part I

Overview of Section 1

Types of Muscle – I

Types of Muscle – II (Illustrations)

Structure of Skeletal Muscle (Illustrations)

Skeletal Muscle Structures and Their Functions

How Skeletal Muscles Contract – The Sliding Filament Theory (incl. Illustration)

The Muscular System – Part I – Summary

Quiz – The Muscular System – Part I

Types of Muscle – I

There are three types of muscle (muscle tissue):

A) Skeletal muscle (or striated muscle)
[Note: Striated = "striped"]
- Also called voluntary muscle;
- Contraction causes movements of the skeleton/skeletal parts;
- Mostly under voluntary control (except for reflexes).

B) Smooth muscle
- Muscle making up the walls of internal organs, eg. blood vessels, the oesophagus (the food tract), the stomach, the intestines;
- Contractions enable movement that aids the functioning of the organs, eg. movement of food from the mouth to the stomach, churning of food in the stomach to aid digestion, movement of nutrients and waste material through the intestines, movement of blood along arteries;
- Under unconscious (involuntary) control by the brain and nervous system.

C) Cardiac muscle
- Muscle making up the walls of the heart;
- Contractions are responsible for the heart beat/pulse and pumping blood (containing vital oxygen) around the body;
- Under unconscious (involuntary) control by the brain and nervous system.

Types of Muscle – II

The three types of muscle (muscle tissue):

A) Skeletal muscle (or striated muscle)
B) Smooth muscle
C) Cardiac muscle

A

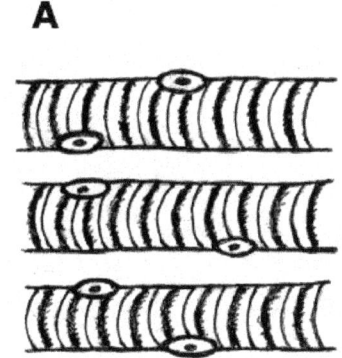

B

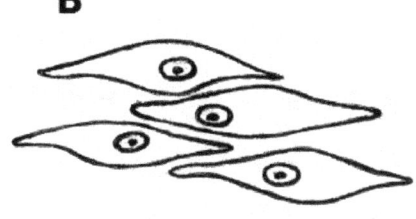

C

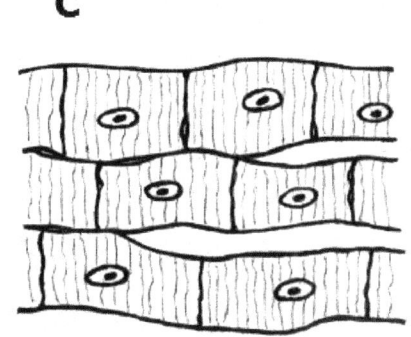

Structure of Skeletal Muscle

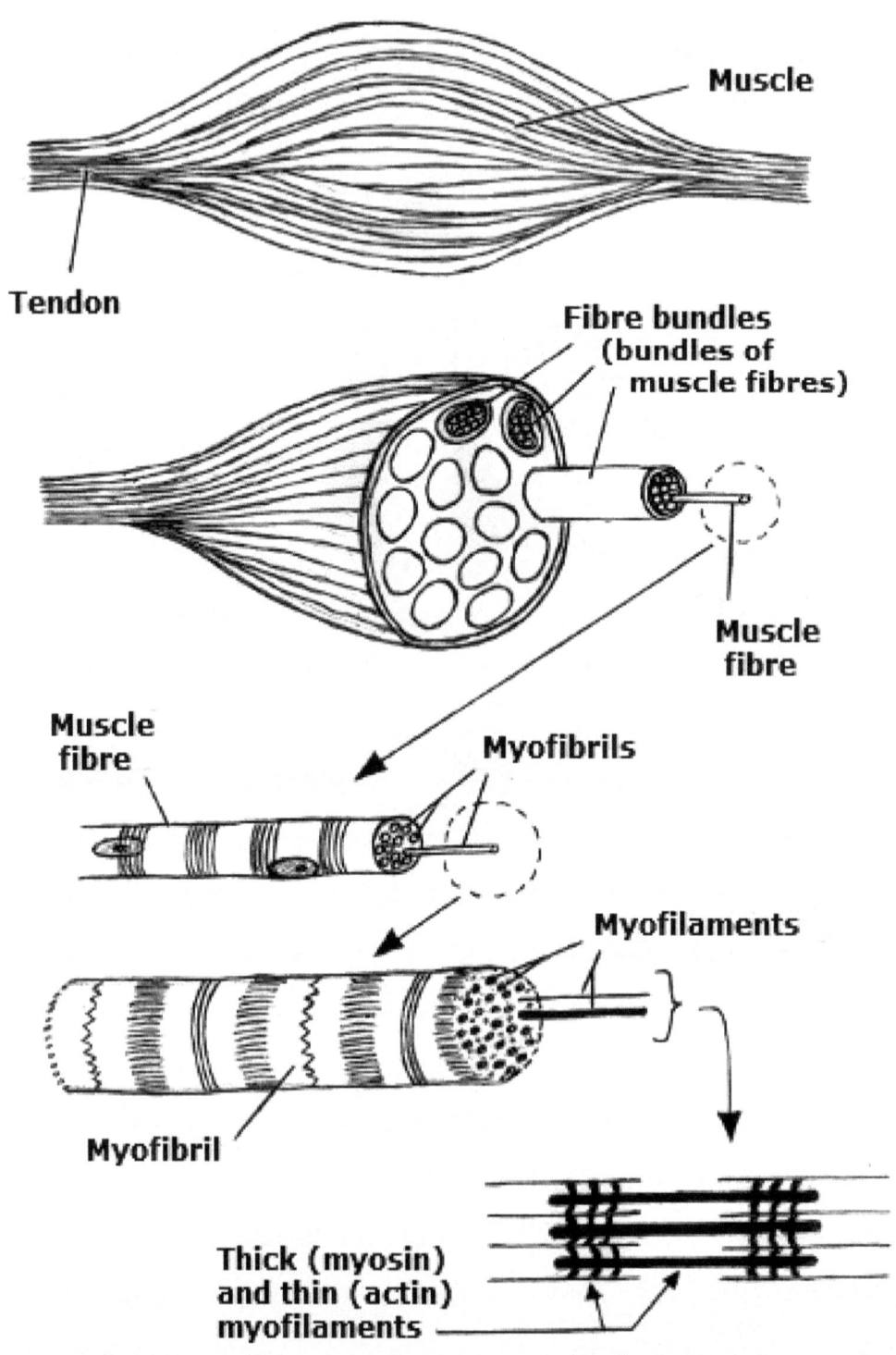

Skeletal Muscle Structures and Their Functions

Muscle (skeletal muscle)
- Composed of bundles of muscle fibres;
- Muscle tissue has a striped ("striated") appearance under the microscope;
- Muscle contraction generally causes movement of a bone about a joint.

Tendon
- Fibrous tissue that attaches muscle to bone.

Muscle fibre
- An individual muscle cell;
- Has a striped appearance;
- Has several nuclei;
- Composed of myofibrils;
- Receives signals from the brain and nervous system via motor nerve fibres, causing it to contract (shorten).

Myofibril
- Composed of myofilaments, of which there are two types: myosin (thick filaments) and actin (thin filaments).

Myofilaments (myosin and actin filaments)
- Microscopic elements responsible for muscle contraction (shortening);
- Also responsible for the striped appearance of skeletal muscle tissue;
- Arranged in parallel rows;
- The thick (myosin) and thin (actin) filaments are able to slide over each other during muscle contraction and hold their position owing to "cross-bridges" on the myosin filaments (which attach to the actin filaments).

Motor nerve fibres
- Branches of motor nerves (from the brain and nervous system) that connect with individual muscle fibres;
- Nerve signals from motor nerve fibres stimulate the muscle fibres (and thus the whole muscle) to contract.

How Skeletal Muscles Contract
– The Sliding Filament Theory –

Background – Structure of Skeletal Muscle

Muscles are composed of bundles of muscle fibres. Each muscle fibre is made up of many tiny myofibrils (a few hundred to several thousand, depending on the width of the muscle fibre). The myofibrils, in turn, are made up of bundles of microscopic myofilaments. There are two types of myofilaments – actin filaments and myosin filaments – which are thought to be responsible for the contractile properties of muscles. Under the microscope, actin filaments appear thin and pale-coloured, whereas myosin filaments are thicker and darker. These colour differences are responsible for the "striated" (striped) appearance of skeletal muscle tissue under the microscope (skeletal muscle is also sometimes called striated muscle).

Sliding Filament Theory*

The actin and myosin filaments are arranged parallel to each other, and the myosin filaments have protrusions on them that are able to link up with binding sites on neighbouring actin filaments to form cross-bridges between the two types of filaments. Nerve impulses from motor nerve fibres connecting to the muscle fibres release chemicals (neurotransmitters) that cause the cross-bridges to pull (or slide) the actin and myosin filaments over each other. This has the effect of shortening the myofibrils (made of the myofilaments), which shortens the muscle fibres, and progressively, the whole muscle (ie. it contracts). When nerve impulses to the muscle fibres stop, or lessen, the actin and myosin filaments are able to slide apart, allowing the muscle to lengthen again and relax.

*Subsequent research has shown the basics of this theory to be essentially correct, although the actual process (now frequently referred to as "excitation-contraction coupling") is somewhat more complex than the above description.

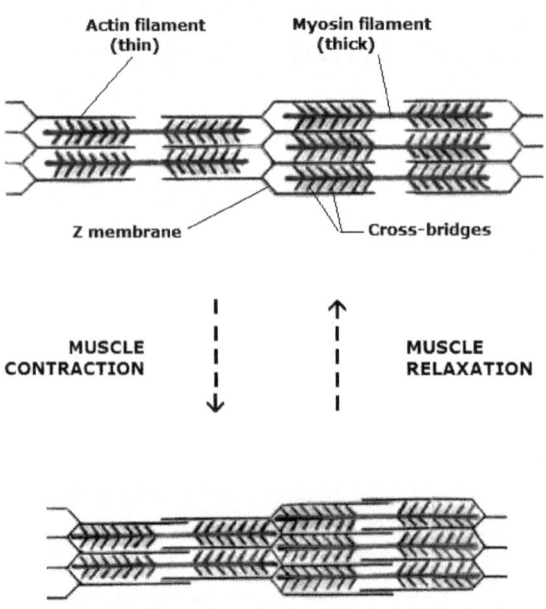

The Muscular System – Part I – Summary

Types of Muscle

There are three types of muscle (muscle tissue):
A) Skeletal muscle (or striated muscle) (striated = "striped")
- Also called voluntary muscle;
- Contraction causes movements of the skeleton/skeletal parts;
- Mostly under voluntary control (except for reflexes).

B) Smooth muscle
- Muscle making up the walls of internal organs, eg. blood vessels, the oesophagus (the food tract), the stomach, the intestines;
- Contractions enable movement that aids the functioning of the organs, eg. movement of food from the mouth to the stomach, churning of food in the stomach to aid digestion, movement of nutrients and waste material through the intestines, movement of blood along arteries;
- Under unconscious (involuntary) control by the brain and nervous system.

C) Cardiac muscle
- Muscle making up the walls of the heart;
- Contractions are responsible for the heart beat/pulse and pumping blood (containing vital oxygen) around the body;
- Under unconscious (involuntary) control by the brain and nervous system.

Skeletal Muscle Structures and Their Functions

Muscle (skeletal muscle)
- Composed of bundles of muscle fibres;
- Muscle tissue has a striped ("striated") appearance under the microscope;
- Muscle contraction generally causes movement of a bone about a joint.

Tendon
- Fibrous tissue that attaches muscle to bone.

Muscle fibre
- An individual muscle cell;
- Has a striped appearance;
- Has several nuclei;
- Composed of myofibrils;
- Receives signals from the brain and nervous system via motor nerve fibres, causing it to contract (shorten).

Myofibril
- Composed of myofilaments, of which there are two types: myosin (thick filaments) and actin (thin filaments).

Myofilaments (myosin and actin filaments)
- Microscopic elements responsible for muscle contraction (shortening);
- Also responsible for the striped appearance of skeletal muscle tissue;
- Arranged in parallel rows;
- The thick (myosin) and thin (actin) filaments are able to slide over each other during muscle contraction and hold their position owing to "cross-bridges" on the myosin filaments (which attach to the actin filaments).

Motor nerve fibres
- Branches of motor nerves (from the brain and nervous system) that connect with individual muscle fibres;
- Nerve signals from the motor nerve fibres stimulate the muscle fibres (and thus the whole muscle) to contract.

How Skeletal Muscles Contract – Sliding Filament Theory

Background – Structure of Skeletal Muscle
Muscles are composed of muscle fibres; muscle fibres are made up of myofibrils; myofibrils are made up of myofilaments; the two types of myofilaments are actin filaments (thin) and myosin filaments (thick); these myofilaments are thought to be responsible for the contractile properties of muscles.

Sliding Filament Theory*
The actin and myosin filaments are arranged parallel to each other. The myosin filaments have protrusions on them that link up with binding sites on neighbouring actin filaments, forming cross-bridges. Motor nerve fibres send signals to the muscle fibres that cause the cross-bridges to pull (slide) the actin and myosin filaments over each other. This shortens the myofibrils (made of myofilaments), which shortens the muscle fibres, which shortens (contracts) the whole muscle. When nerve signals to the muscle fibres stop, the actin and myosin filaments can slide apart again, allowing the muscle to relax.
[*AKA "excitation-contraction coupling".]

Quiz – The Muscular System – Part I

1. Name the three types of muscles (muscle tissue): .... ,

.... , and ....

2. .... muscles are largely under voluntary control, whereas

.... muscles and .... muscles are

under unconscious (involuntary) control by the brain and nervous system.

3. An individual muscle cell is called a ....

4. Muscles are made up of bundles of .... , which in turn are

made up of .... , which in turn are made up of

....

5. There are two types of myofilaments; these are called .... and

.... filaments.

6. The structures made of fibrous connective tissue that attach muscles to bone are called

....

7. Nerve impulses travelling along motor nerve fibres release chemicals in the region of the

muscle fibres that cause a muscle to ....

8. The microscopic elements of skeletal muscle that are thought to be responsible for its

.... properties, and also its striped appearance under a

microscope, are called .... and ....

filaments.

* * *

Section 2

The Muscular System – Part II

Overview of Section 2

Anatomical Positions – I

Anatomical Positions – II (Illustration)

Terms Relating to Muscle Actions/Movements – I

Terms Relating to Muscle Actions/Movements – II (Illustrations)

Terms Relating to Muscle Actions/Movements – III (Illustration)

Terms Relating to Muscle Actions/Movements – IV (Illustration)

Names and Functions of the Some of the Main Muscles of the Body – I
- Front of the Body (Neck and Trunk); Back of the Body (Neck and Trunk)

Muscles of the Front of the Body – Neck and Trunk (Illustration)

Muscles of the Back of the Body – Neck and Trunk (Illustration)

Names and Functions of the Some of the Main Muscles of the Body – II
- Front of the Arm; Back of the Arm; Front of the Leg; Back of the Leg

Muscles of the Front of the Arm (Illustration)

Muscles of the Back of the Arm (Illustration)

Muscles of the Front of the Leg (Illustration)

Muscles of the Back of the Leg (Illustration)

The Muscular System – Part II – Summary

Quiz – The Muscular System – Part II

The Muscular System – Part II

The contents of Part II of The Muscular System refer to the skeletal muscles of the body – ie. those muscles involved largely in voluntary movements.

Anatomical Positions – I

Terms that refer to anatomical positions are used to describe not only many skeletal muscles, but also bones, nerves, blood vessels and other structures of the body. Below are the commonly used terms and their opposites, and their meanings.

Superior
- Above (towards the head)

Inferior (opposite term)
- Below (towards the feet)

Lateral
- Towards the side of the body

Medial (opposite term)
- Towards the centre (midline) of the body

Anterior (also Ventral)
- Towards the front of the body

Posterior (also Dorsal) (opposite terms)
- Towards the back of the body

Proximal
- Nearer to the heart (tends to refer to a limb or part of a limb)

Distal (opposite term)
- Further away from the heart (tends to refer to a limb or part of a limb)

Anatomical Positions – II

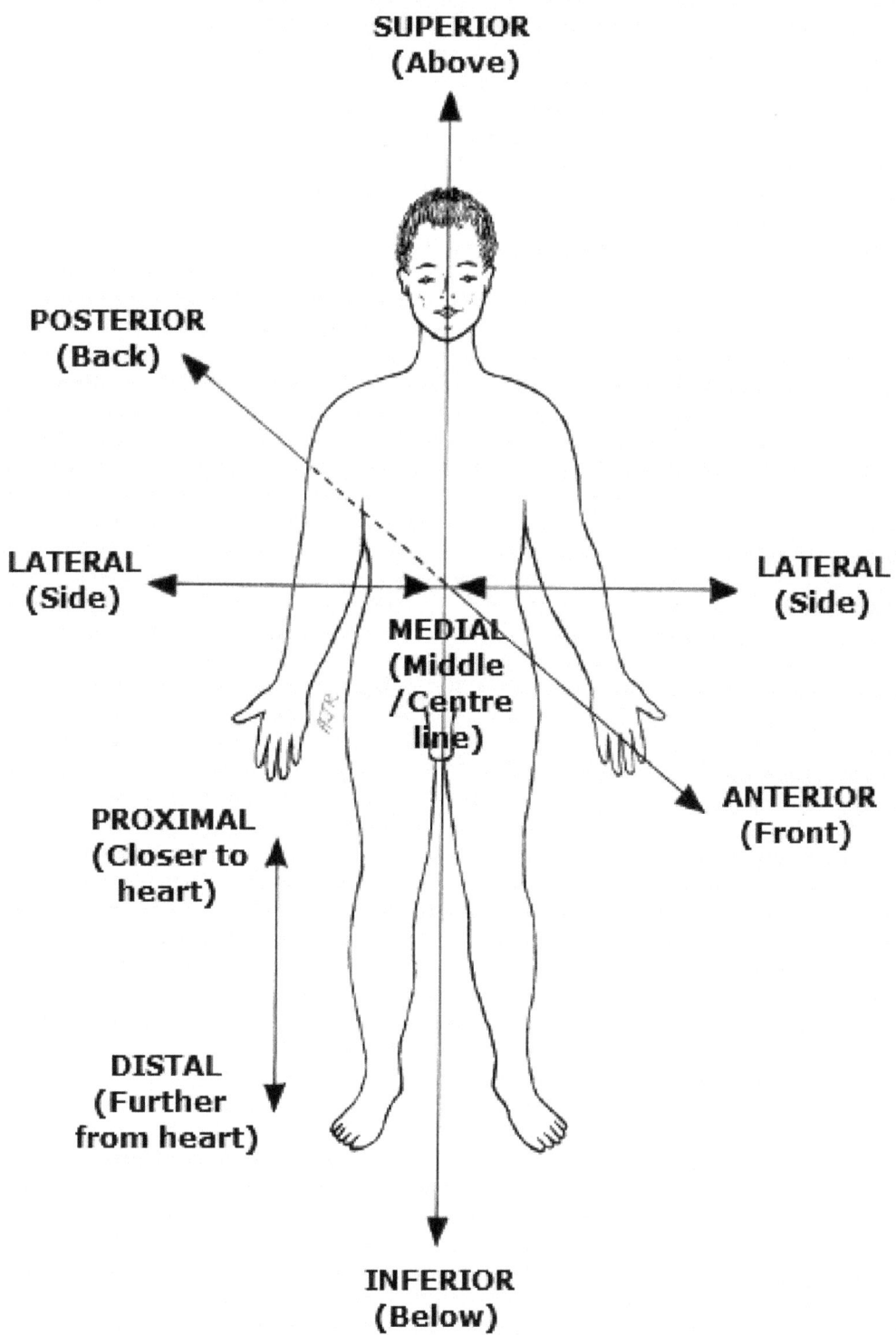

Terms Relating to Muscle Actions/Movements – I
(also Movements/Positions of a Body Part or the Whole Body)

Below are the commonly used terms and their opposites, and their meanings.

Flexion
- muscle contraction usually **flexes (bends)** a limb

Extension (opposite term)
- muscle contraction usually **extends (straightens)** a limb

Adduction
- muscle contraction usually moves a limb (or other body part) **towards** the body

Abduction (opposite term)
- muscle contraction usually moves a limb (or other body part) **away from** the body

Dorsiflexion (of the foot)
- muscle contraction moves the foot **upwards** at the ankle

Plantarflexion (of the foot)
- muscle contraction moves the foot **downwards** at the ankle

Inversion (of the foot)
- muscle contraction moves the sole of the foot **inwards**

Eversion (of the foot)
- muscle contraction moves the sole of the foot **outwards**

Pronation
- Palm-down or sole-down (referring to a limb)

Supination (opposite term)
- Palm-up or sole-up (referring to a limb)

Prone
- Lying face-down (referring to the whole body)

Supine (opposite term)
- Lying face-up (referring to the whole body)

Terms Relating to Muscle Actions/Movements – II
(also Movements/Positions of a Body Part or the Whole Body)

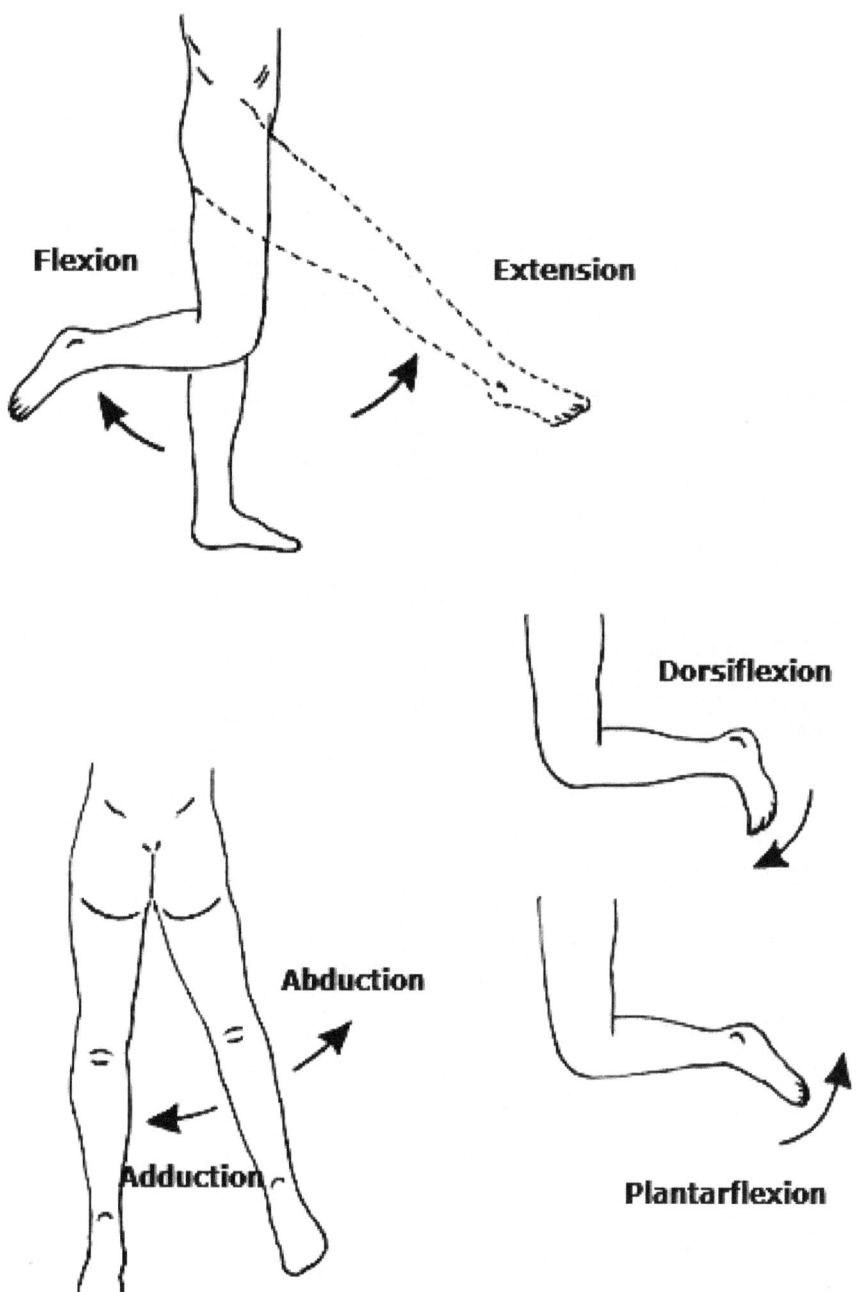

Terms Relating to Muscle Actions/Movements – III
(also Movements/Positions of a Body Part or the Whole Body)

Right arm

Supination (Palm up)

Pronation (Palm down)

Terms Relating to Muscle Actions/Movements – IV
(also Movements/Positions of a Body Part or the Whole Body)

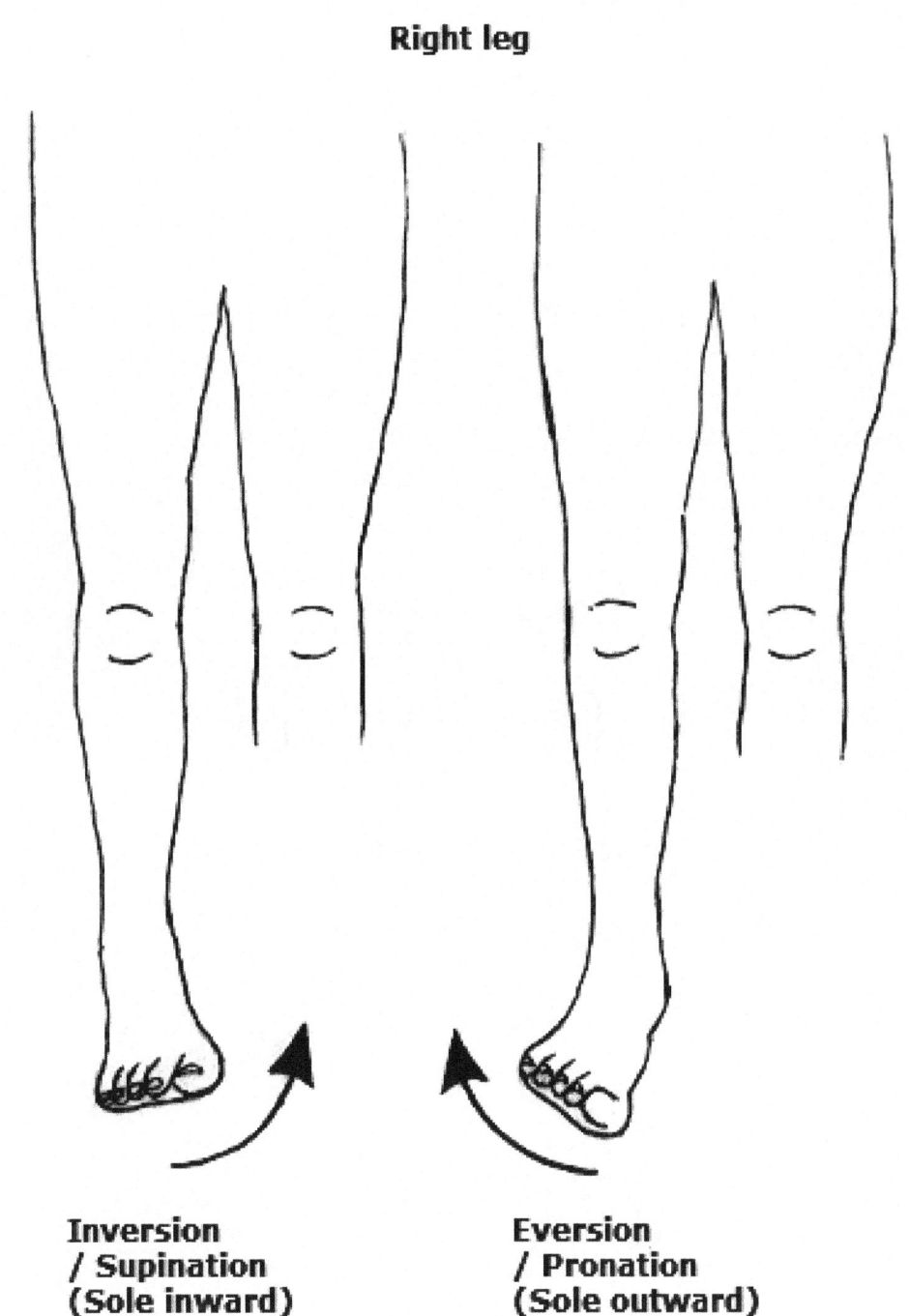

Names and Functions of the Some of the Main Muscles of the Body – I

– Front of the Body – Neck and Trunk –

Neck:
- **Sternocleidomastoid:** Aids in turning movements of the neck; bends (flexes) the head towards the neck/shoulder on one or other side; turns the neck/head from side to side.

Trunk:
- **Pectoralis major (chest):** Helps to move the shoulder and the upper arm (downwards).
- **Transverse abdominis (abdomen):** Supports the abdominal organs; aids an active out-breath (draws the abdominal muscles inwards).
- **Rectus abdominis (abdomen):** Supports the abdominal organs; flexes the vertebral column (spine).

– Back of the Body – Neck and Trunk –

Neck:
- **Splenius capitis:** Holds the head upright/returns it to an upright position.

Trunk:
- **Trapezius (neck, shoulders, upper back):** Rotates the shoulder; raises the shoulder; pulls the shoulder backwards.
- **Latissimus dorsi (middle back):** Pulls the shoulder in towards the body (adducts); extends (pulls down) the upper arm.
- **Gluteus maximus (buttocks):** Abducts the femur; extends the leg.

Muscles of the Front of the Body – Neck and Trunk

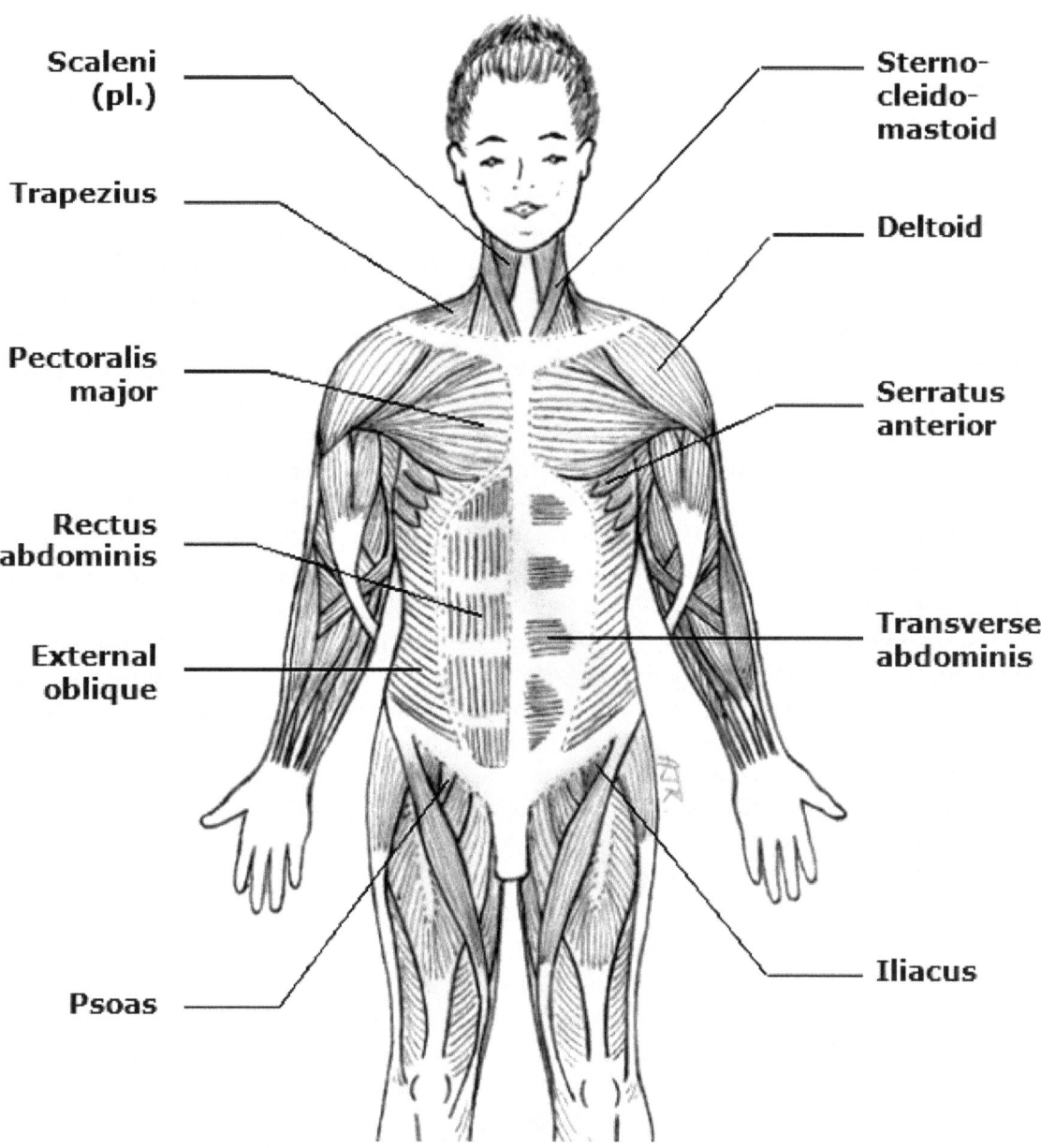

Muscles of the Back of the Body – Neck and Trunk

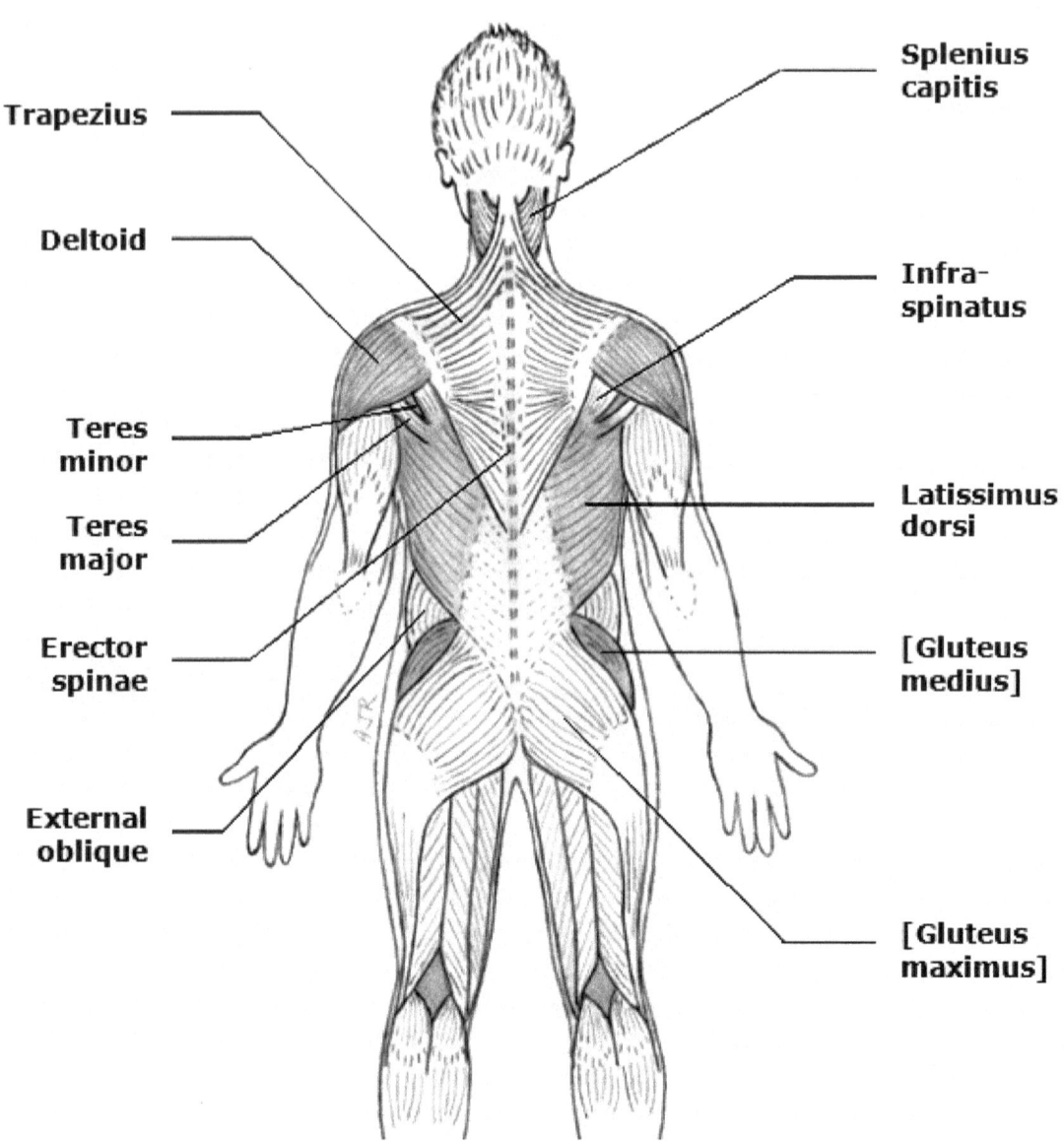

[ ] = Muscles of the hip/lower extremity

Names and Functions of the Some of the Main Muscles of the Body – II

– Front of the Arm –

- **Deltoid (shoulder/upper arm):** Helps to move the shoulder and upper arm.
- **Biceps (upper arm – flexion):** Bends (flexes) and helps turn the forearm.

– Back of the Arm –

- **Triceps (upper arm – extension):** Straightens (extends) the forearm.

– Front of the Leg –

- **Rectus femoris (upper leg) (part of the quadriceps*):** Raises (flexes) the thigh; straightens (extends) the leg at the knee. [*Quadriceps = Rectus femoris + Vastus intermedius + Vastus medialis + Vastus lateralis.]
- **Tibialis anterior (lower leg):** Bends (flexes) the foot upwards (dorsiflexion).

– Back of the Leg –

- **Biceps femoris (upper leg) (one of the hamstrings**):** Lowers (extends) the thigh; bends (flexes) the leg at the knee. [**Hamstrings = Biceps femoris + Semitendinosus + Semimembranosus.]
- **Gastrocnemious (lower leg):** Bends (flexes) the knee; points the foot downwards at the ankle (plantarflexion of the foot).
- **Soleus (lower leg):** Points the foot downwards at the ankle (plantarflexion of the foot).

Muscles of the Front of the Arm

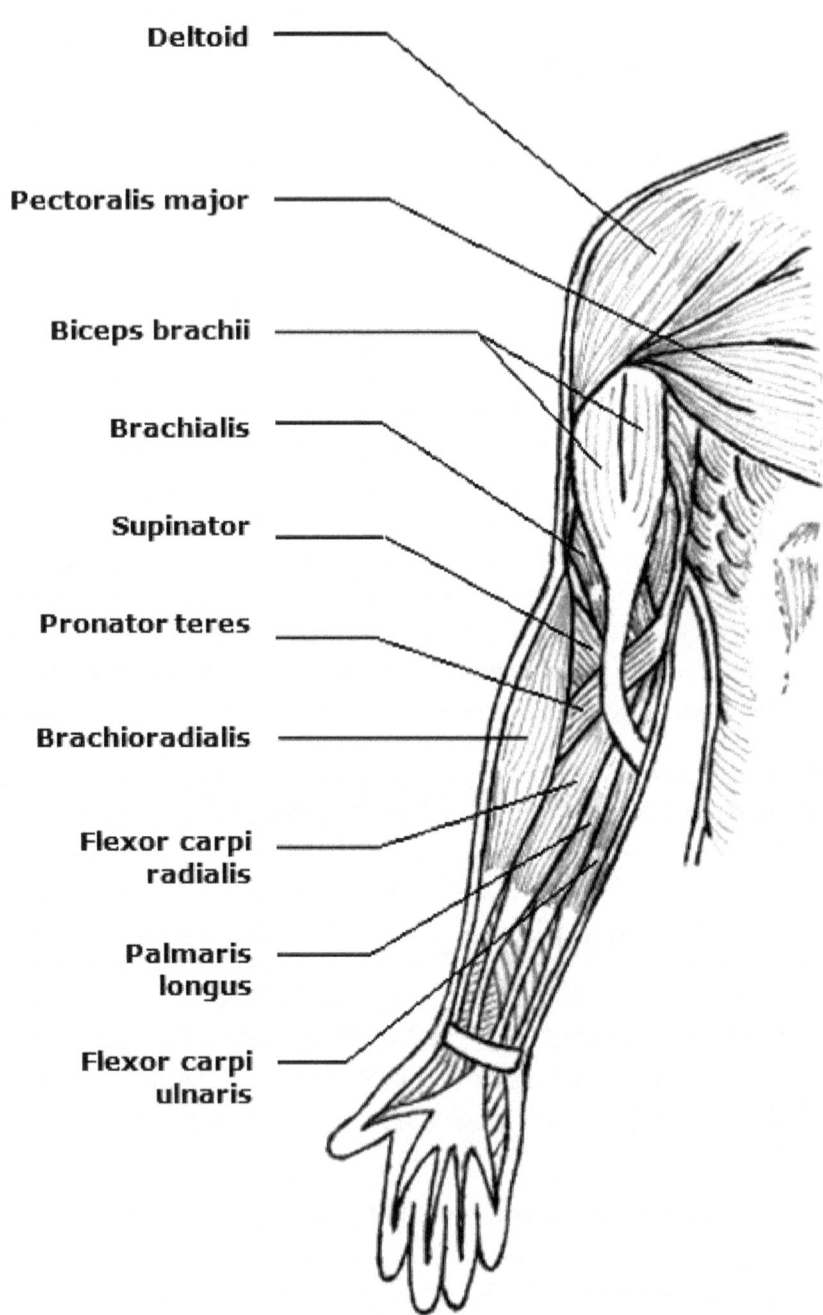

Muscles of the Back of the Arm

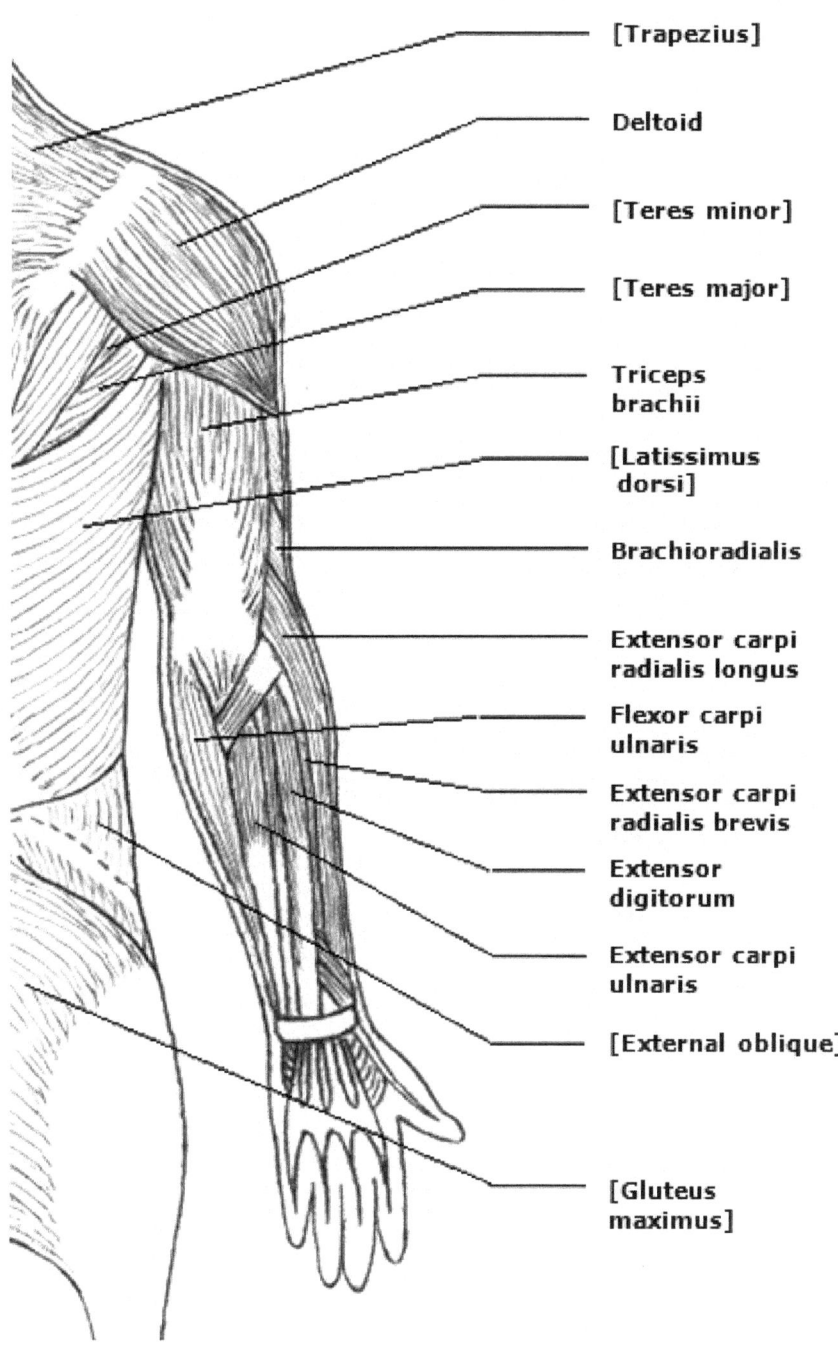

Note: Some muscles of the torso or hip are indicated in brackets.

Muscles of the Front of the Leg

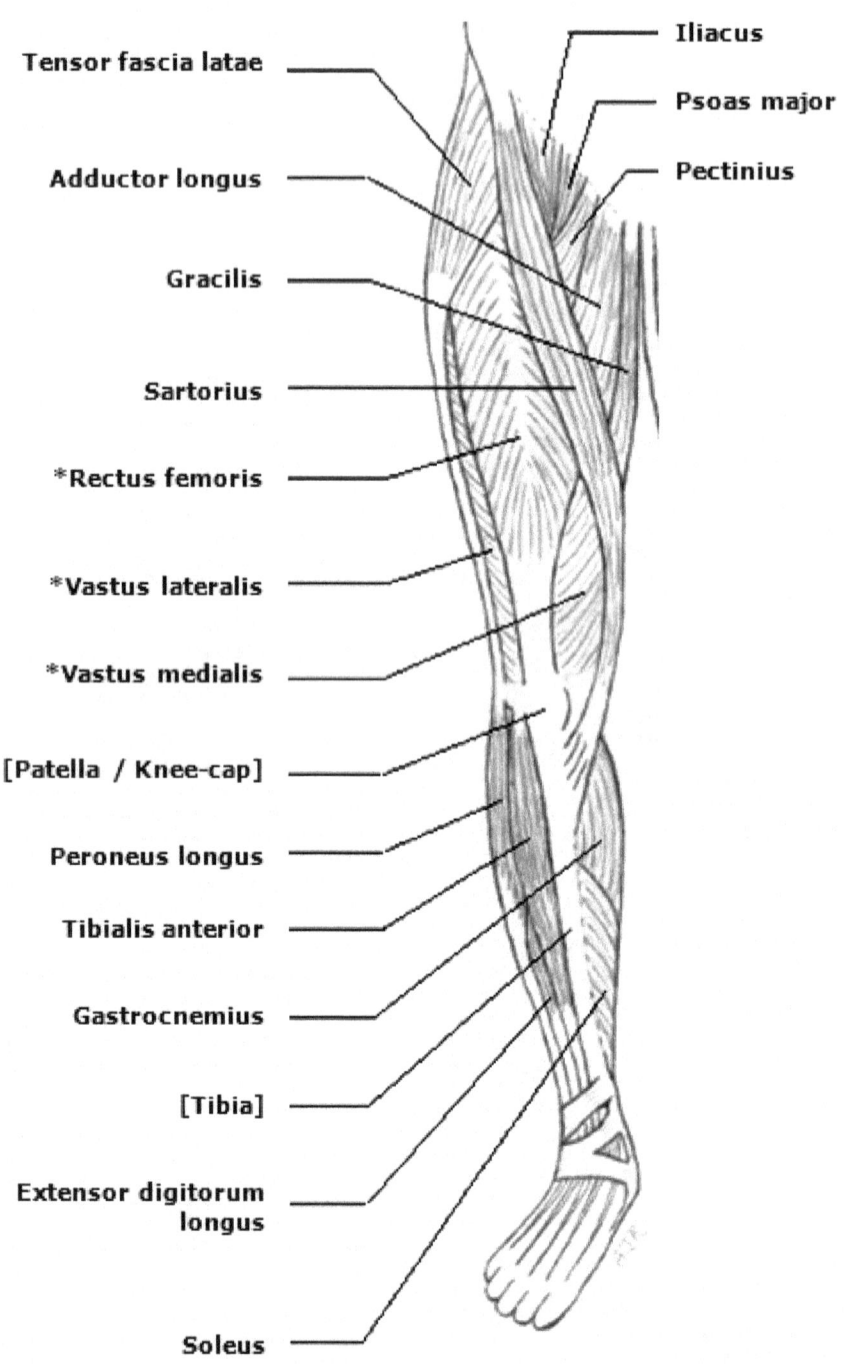

* + Vastus intermedius = Quadriceps [ ] = Bone

Muscles of the Back of the Leg

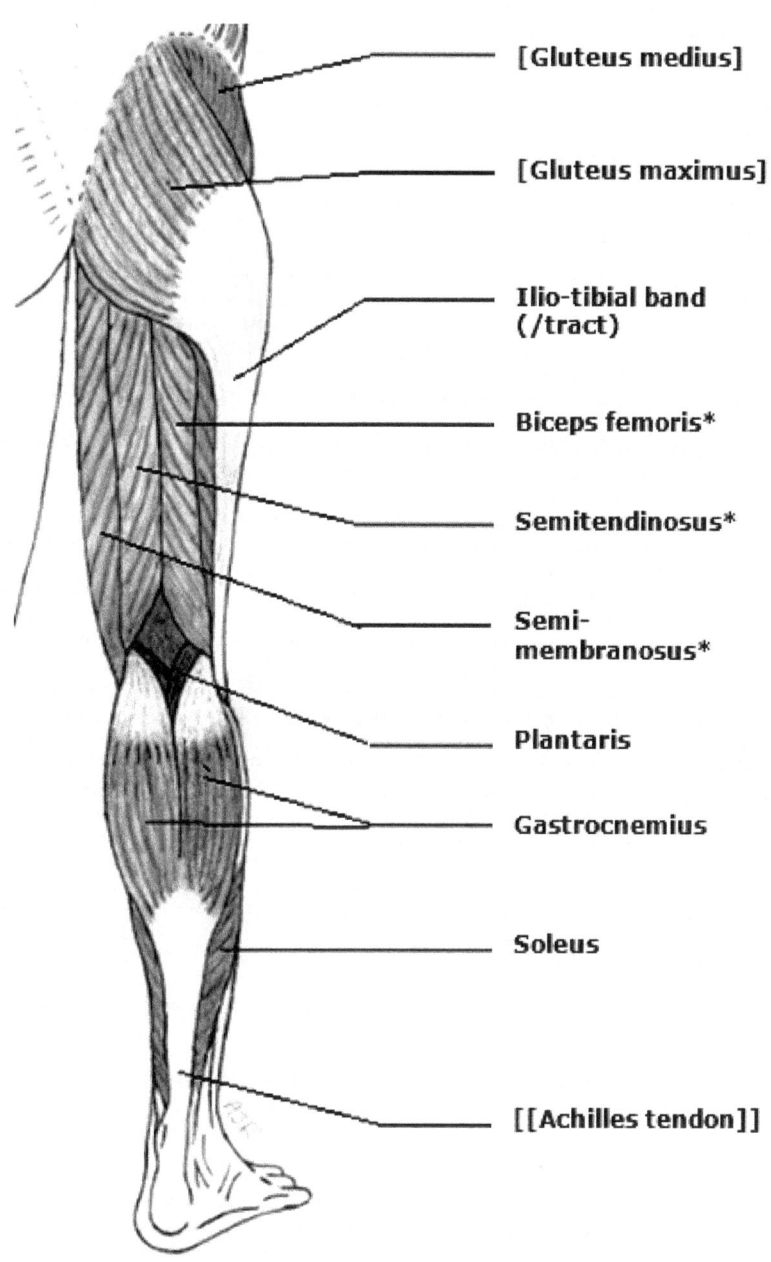

* = Hamstrings

[ ] = Muscles of the hip
[[ ]] = Tendon

The Muscular System – Part II – Summary

Anatomical Positions

Terms that refer to anatomical positions are used to describe not only many skeletal muscles, but also bones, nerves, blood vessels and other structures of the body.

Superior – Above (towards the head)
Inferior (opposite term) – Below (towards the feet)
Lateral – Towards the side of the body
Medial (opposite term) – Towards the centre (midline) of the body
Anterior (also Ventral) – Towards the front of the body
Posterior (also Dorsal) (opposite terms) – Towards the back of the body
Proximal – Nearer to the heart (tends to refer to a limb or part of a limb)
Distal (opposite term) – Further away from the heart (tends to refer to a limb or part of a limb)

Some Terms Relating to Muscle Actions/Movements
(also Movements/Positions of a Body Part or the Whole Body)

Flexion – muscle contraction usually flexes (bends) a limb
Extension (opposite term) – muscle contraction usually extends (straightens) a limb
Adduction – muscle contraction usually moves a limb (or other body part) towards the body
Abduction (opposite term) – muscle contraction usually moves a limb (or other body part) away from the body
Dorsiflexion (of the foot) – muscle contraction moves the foot upwards at the ankle
Plantarflexion (of the foot) – muscle contraction moves the foot downwards at the ankle
Inversion (of the foot) – muscle contraction moves the sole of the foot inwards
Eversion (of the foot) – muscle contraction moves the sole of the foot outwards
Pronation – Palm-down or sole-down (referring to a limb)
Supination (opposite term) – Palm-up or sole-up (referring to a limb)
Prone – Lying face-down (referring to the whole body)
Supine (opposite term) – Lying face-up (referring to the whole body)

Names and Functions of the Some of the Main Muscles of the Body – I

Front of the Body (Neck and Trunk):
  Neck:
- **Sternocleidomastoid:** Aids in turning movements of the neck; bends (flexes) the head towards the neck/shoulder on one or other side; turns the neck/head from side to side.
  Trunk:
- **Pectoralis major (chest):** Helps to move the shoulder and the upper arm (downwards).
- **Transverse abdominis (abdomen):** Supports the abdominal organs; aids an active out-breath (draws the abdominal muscles inwards).
- **Rectus abdominis (abdomen):** Supports the abdominal organs; flexes the vertebral column (spine).

Names and Functions of the Some of the Main Muscles of the Body – I
(continued)

Back of the Body (Neck and Trunk):
  Neck:
- **Splenius capitis:** Holds the head upright / returns it to an upright position.
  Trunk:
- **Trapezius (neck, shoulders, upper back):** Rotates the shoulder; raises the shoulder; pulls the shoulder backwards.
- **Latissimus dorsi (middle back):** Pulls the shoulder in towards the body (adducts); extends (pulls down) the upper arm.
- **Gluteus maximus (buttocks):** Abducts the femur; extends the leg.

Names and Functions of the Some of the Main Muscles of the Body – II

Front of the Arm:
- **Deltoid (shoulder/upper arm):** Helps to move the shoulder and upper arm.
- **Biceps (upper arm – flexion):** Bends (flexes) and helps turn the forearm.

Back of the Arm:
- **Triceps (upper arm – extension):** Straightens (extends) the forearm.

Front of the Leg:
- **Rectus femoris (upper leg) (part of the quadriceps*):** Raises (flexes) the thigh; straightens (extends) the leg at the knee. [*Quadriceps = Rectus femoris + Vastus intermedius + Vastus medialis + Vastus lateralis.]
- **Tibialis anterior (lower leg):** Bends (flexes) the foot upwards (dorsiflexion).

Back of the Leg
- **Biceps femoris (upper leg) (one of the hamstrings**):** Lowers (extends) the thigh; bends (flexes) the leg at the knee. [**Hamstrings = Biceps femoris + Semitendinosus + Semimembranosus.]
- **Gastrocnemious (lower leg):** Bends (flexes) the knee; points the foot downwards at the ankle (plantarflexion of the foot).
- **Soleus (lower leg):** Points the foot downwards at the ankle (plantarflexion of the foot).

Quiz – The Muscular System – Part II

1. When muscle contraction causes a limb (arm or leg) to bend, the action is called ....

2. When muscle contraction causes a limb (or other body part) to move away from the body, the action is called ....

3. The large muscles across the chest are called the ....

4. The large muscle covering the top of the shoulder, which helps move the shoulder and upper arm, is the .... muscle.

5. Anatomical structures located close (or closer) to the heart are termed .... (in terms of their anatomical position).

6. The three-headed muscle at the back of the upper arm is the ....

7. The rectus femoris muscle is one of .... [number] muscles on the front of the thigh that together make up the .... muscles.

8. The large muscle extending from the nape of the neck, across the shoulder and down to the upper mid-back is called the ....

9. The main large muscle of the buttock is known as the ....

10. The semitendinosus muscle at the back of the thigh is one of the ....

11. The main bifurcated muscle of the calf (the back of the lower leg) is the .... muscle.

12. The tendon attaching the above mentioned muscle to the heel is the .... tendon.

Section 3

The Cardiovascular System – Part I

Overview of Section 3

The Cardiovascular System

What is the Cardiovascular System?

Functions of the Cardiovascular System

Composition of Blood and Functions of Blood Components

Types and Characteristics of Blood Vessels

Circulation of Blood – Schematic (Diagram)

Two Main (Blood) Circulations – Schematic (Diagrams)
- Systemic Circulation (Heart – Body – Heart)
- Pulmonary Circulation (Heart – Lungs – Heart)

The Cardiovascular System – Part I – Summary

Quiz – The Cardiovascular System – Part I

The Cardiovascular System

Also known as:
- The Vascular System
- The Circulatory System
- The Heart and Blood Circulation

What is the Cardiovascular System?

The cardiovascular system comprises:
- The Blood
- The Heart
- Arteries
- Veins
- Smaller Blood Vessels (Arterioles, Venules and Capillaries)

Functions of the Cardiovascular System

Transport of essential substances around the body
Includes transport of oxygen, nutrients, hormones, enzymes, antibodies and white blood cells.

Transport of waste materials to organs that eliminate them from the body
Substances transported include urea (to the kidneys) and carbon dioxide (to the lungs).

Helps to fight infection
Via antibodies and white blood cells.

Prevents loss of body fluids after injury
Via blood clotting function.

Regulation of body temperature
By redirecting blood flow towards or away from the surface of the body.

Composition of Blood and Functions of Blood Components

Blood is made up of:

1. Plasma (55%)

Properties:

- Straw-coloured fluid made up of water (approx. 92%) and plasma proteins.

Functions:

- To transport blood cells plus mineral salts, nutrients, waste products, hormones, enzymes, dissolved gases (eg. carbon dioxide), antibodies and antitoxins.

2. Blood Cells (45%)

A) Red Blood Cells (Erythrocytes)

Properties:

- Make up 99% of the blood cell population;
- Have no nucleus;
- Contain haemoglobin, which binds oxygen;
- Live for about 120 days.

Functions:

- To transport oxygen around the body.

B) White Blood Cells (Leukocytes)

Properties:

- Make up 0.2% of the blood cell population;
- Include granulocytes, lymphocytes and monocytes;
- Larger than red blood cells;
- Have a nucleus.

Functions:

- To help protect against bacteria and viruses;
- Actions include production of antibodies and engulfing and destroying bacteria.

C) Platelets (Thrombocytes)

Properties:

- Make up 0.8% of the blood cell population;
- Tiny blood cells with no nucleus.

Functions:

- To activate clotting factors in the blood in response to injury.

Types and Characteristics of Blood Vessels

Arteries
Vessels walls – Thick, muscular, elastic;
Transport of blood – Away from the heart;
Blood flow – Fast and pulsatile; pumped by the heart and contractions of the (smooth) muscles in the artery walls;
Blood pressure – Relatively high;
Blood – Contains oxygen (oxygenated blood) (except the pulmonary arteries); contains nutrients.

Veins
Vessels walls – Thin;
Transport of blood – Towards the heart;
Blood flow – Slow; pumped by the action/contractions of skeletal muscles; back-flow of blood is prevented by valves on the inside walls of the veins;
Blood pressure – Low;
Blood – Contains carbon dioxide (deoxygenated blood) (except the pulmonary veins); contains waste products.

Arterioles
- Smaller versions (branches) of arteries;
- Walls are muscular, enabling the arterioles to constrict or dilate to carry less or more blood to the cells, as required;
- At the level of the body cells, they branch to form capillaries, through which nutrients and oxygen are transported to the body cells.

Venules
- Smaller versions (branches) of veins;
- Walls are thin and easily collapsed (which can lead to build up of fluid [oedema] in the tissues);
- At the level of the body cells, they connect to the capillaries, from which they collect carbon dioxide and waste products from the body cells.

Capillaries
- The smallest blood vessels;
- Walls are one cell thick and porous;
- Nutrients, gases (oxygen and carbon dioxide) and waste products are able to pass between the inside of the capillaries and the body cells.

Circulation of Blood – Schematic

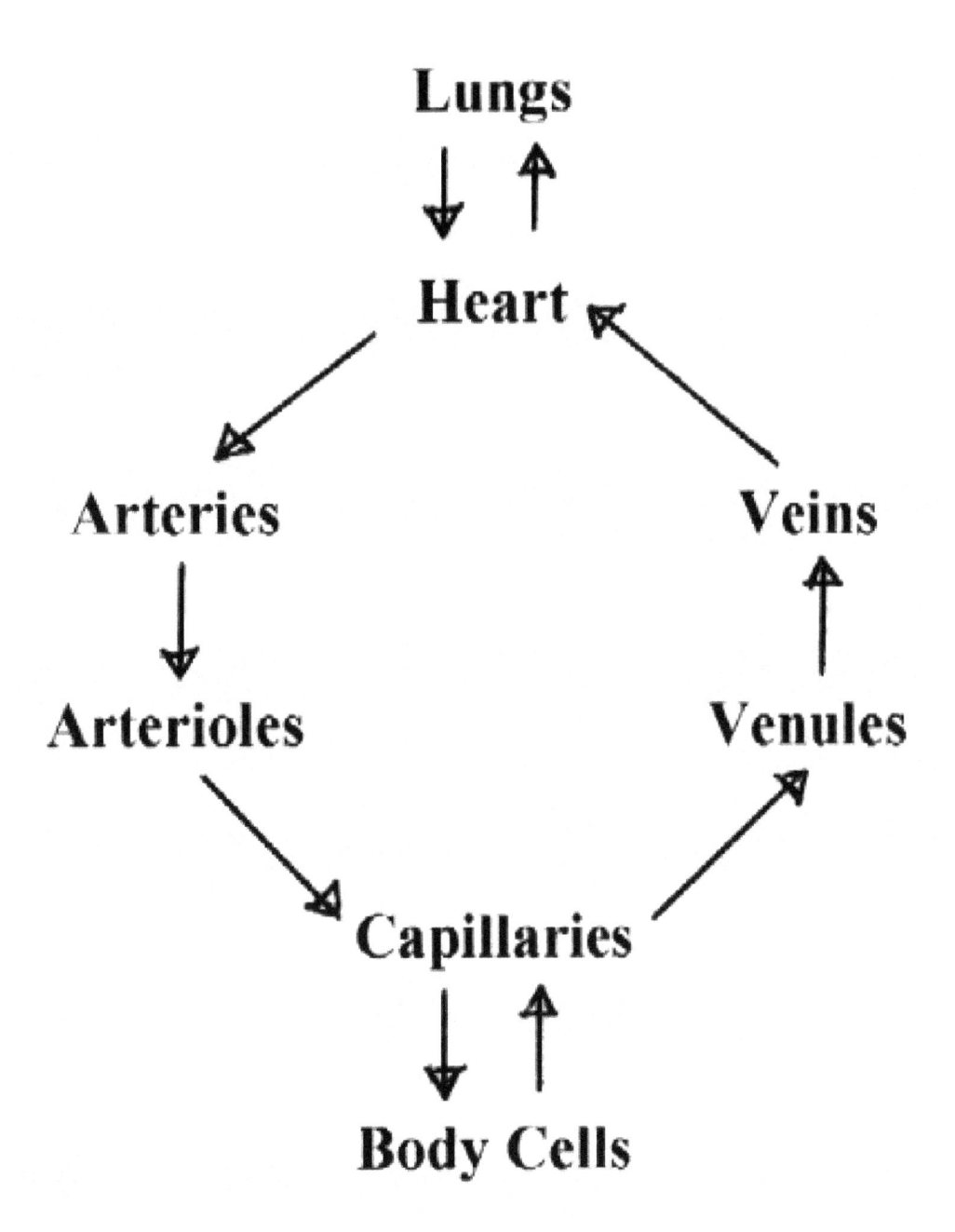

Lungs → Heart → Arteries → Arterioles → Capillaries → Body Cells → Capillaries → Venules → Veins → Heart → Lungs

Two Main (Blood) Circulations – Schematic

1) Systemic Circulation (Heart – Body – Heart)
- Pumps oxygenated blood from the heart, through the arteries and around the body to supply oxygen and nutrients to all cells of the body;
- Returns deoxygenated blood from the body cells to the heart, via the veins.

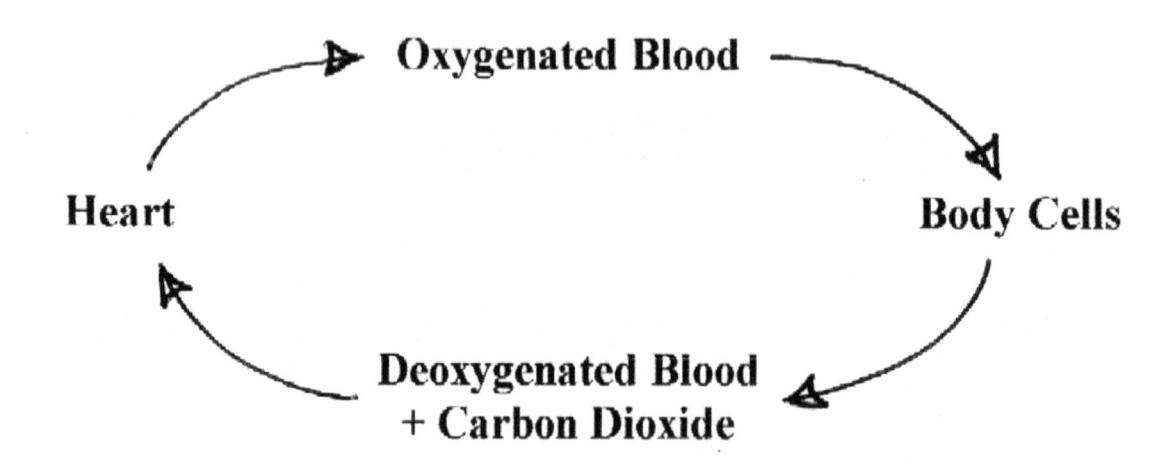

Heart → Oxygenated Blood → Body Cells → Deoxygenated Blood + Carbon Dioxide → Heart

2) Pulmonary Circulation (Heart – Lungs – Heart)
- Pumps deoxygenated blood from the heart to the lungs in order to pick up oxygen (from inhaled air);
- Returns the oxygenated blood to the heart.

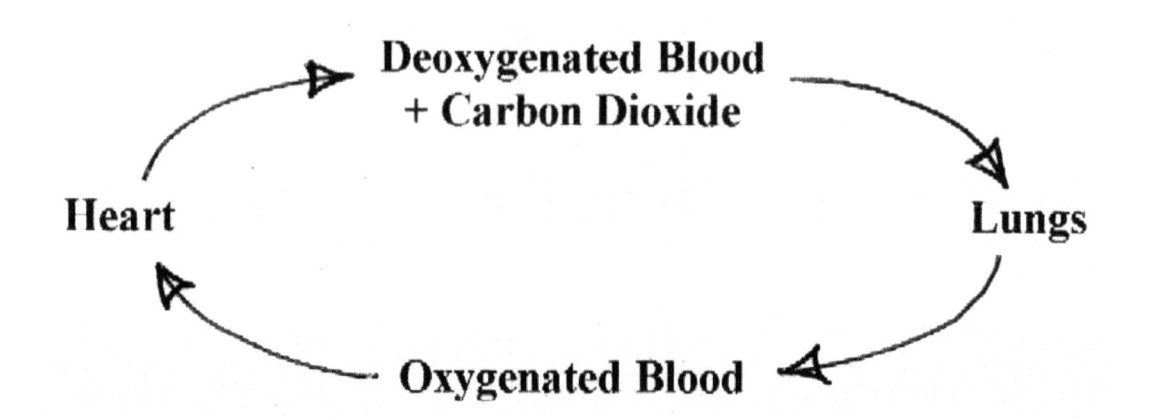

Heart → Deoxygenated Blood + Carbon Dioxide → Lungs → Oxygenated Blood → Heart

The Cardiovascular System – Part I – Summary

Also known as: The Vascular System, The Circulatory System, or The Heart and Blood Circulation.

What is the Cardiovascular System?

The cardiovascular system comprises: The Blood, The Heart, Arteries, Veins and Smaller Blood Vessels (Arterioles, Venules and Capillaries).

Functions of the Cardiovascular System

Transport of essential substances around the body – Including oxygen, nutrients, hormones, enzymes, antibodies, white blood cells.
Transport of waste materials to organs that eliminate them from the body – Including urea (to the kidneys) and carbon dioxide (to the lungs).
Helps to fight infection – Via antibodies and white blood cells.
Prevents loss of body fluids after injury – Via blood clotting function.
Regulation of body temperature – By redirecting blood flow towards or away from the surface of the body.

Composition of Blood and Functions of Blood Components

Blood is made up of :
1. Plasma (55%)
Properties: Straw-coloured fluid made up of water (approx. 92%) and plasma proteins.
Functions: To transport blood cells plus mineral salts, nutrients, waste products, hormones, enzymes, dissolved gases (eg. carbon dioxide), antibodies and antitoxins.
2. Blood Cells (45%)
A) Red Blood Cells (Erythrocytes)
Properties: Make up 99% of the blood cell population; Have no nucleus; Contain haemoglobin, which binds oxygen; Live for about 120 days.
Functions: To transport oxygen around the body.
B) White Blood Cells (Leukocytes)
Properties: Make up 0.2% of the blood cell population; Include granulocytes, lymphocytes and monocytes; Larger than red blood cells; Have a nucleus.
Functions: To help protect against bacteria and viruses; Actions include production of antibodies and engulfing and destroying bacteria.
C) Platelets (Thrombocytes)
Properties: Make up 0.8% of the blood cell population; Tiny blood cells with no nucleus.
Functions: To activate clotting factors in the blood in response to injury.

Characteristics of Blood Vessels

Arteries
- Vessels walls are thick, muscular and elastic; Transport of blood is away from the heart; Blood flow is fast and pulsatile, pumped by the heart and contractions of the (smooth) muscles in the artery walls; Blood pressure is relatively high; Blood contains oxygen (oxygenated blood) (except the pulmonary arteries) and nutrients.

Veins
- Vessels walls are thin; Transport of blood is towards the heart; Blood flow is slow, pumped by the action/contractions of skeletal muscles; Back-flow of blood is prevented by valves on the inside walls of the veins; Blood pressure is low; Blood contains carbon dioxide (deoxygenated blood) (except the pulmonary veins) and waste products.

Arterioles
- Smaller versions (branches) of arteries; Walls are muscular, enabling the arterioles to constrict or dilate to carry less or more blood to the cells, as required; At the level of the body cells, they branch to form capillaries, through which nutrients and oxygen are transported to the body cells.

Venules
- Smaller versions (branches) of veins; Walls are thin and easily collapsed (which can lead to build up of fluid [oedema] in the tissues); At the level of the body cells, they connect to the capillaries, from which they collect carbon dioxide and waste products from the body cells.

Capillaries
- The smallest blood vessels; Walls are one cell thick and porous; Nutrients, gases (oxygen and carbon dioxide) and waste products are able to pass between the inside of the capillaries and the body cells.

Two Main (Blood) Circulations

Systemic Circulation (Heart – Body – Heart)
- Pumps oxygenated blood from the heart, through the arteries and around the body to supply oxygen and nutrients to all cells of the body;
- Returns deoxygenated blood (containing carbon dioxide) from the body cells to the heart, via the veins.

Pulmonary Circulation (Heart – Lungs – Heart)
- Pumps deoxygenated blood from the heart to the lungs in order to pick up oxygen (from inhaled air);
- Returns the oxygenated blood to the heart.

Quiz – The Cardiovascular System – Part I

1. What are the five main functions of the cardiovascular system? .... , ... , ... , ... , ...

2. Blood is made up of plasma, red blood cells, ... and ....

3. Red blood cells transport .... around the body.

4. The type of cells in the blood that are involved in fighting infectious agents such as bacteria and viruses are called ....

5. The type of cells in the blood that are involved in blood clotting are called ....

6. Arteries carry blood .... the heart, whilst veins carry blood .... the heart.

7. Arteries (except the pulmonary arteries) carry blood that has a high content of ....

8. Veins (except the pulmonary veins) carry blood that has a high content of ....

9. The smallest type of blood vessels are called ....

10. The circulatory system that transports blood from the heart to the body cells and back is called the .... circulation.

11. The circulatory system that transports blood from the heart to the lungs and back is called the .... circulation.

* * *

Section 4

The Cardiovascular System – Part II

Overview of Section 4

Structure and Function of the Heart – I

Structure of the Heart – Schematic (Illustration)

Structure and Function of the Heart – II

Heart Contraction (Pumping Action) – I

Heart Contraction (Pumping Action) – II (Illustrations)

The Cardiovascular System – Schematic (Diagram)

The Arterial System – Main Arteries (Illustration)

The Venous System – Main Veins (Illustration)

The Cardiovascular System – Part II – Summary

Quiz – The Cardiovascular System – Part II

Structure and Function of the Heart – I

Four heart chambers

The heart is made up of four chambers: the upper chambers are the right and left atria (singular = atrium); the lower chambers are the right and left ventricles.

[Note: When looking at a diagram of the heart, remember that it is as if you are looking at another person standing in front of you. So, what you see on the left is the right side of the heart, and what you see on the right is the left side of the heart.]

Left side of the heart – Oxygenated blood

The left side of the heart deals with oxygenated blood, which is received from the lungs, via the pulmonary veins, into the left atrium. The blood flows into the left ventricle and is then pumped out through the aorta (the large systemic artery leaving the heart) to supply all areas of the body (via the systemic circulation) (see Section 3: *Two Main (Blood) Circulations*).

Right side of the heart – Deoxygenated blood

The right side of the heart deals with deoxygenated blood from the body cells, which is received via the superior and inferior venae cavae (singular = vena cava) into the right atrium. The blood flows into the right ventricle and is then pumped out through the pulmonary arteries to the lungs (via the pulmonary circulation) (see Section 3: *Two Main (Blood) Circulations*).

Four sets of valves prevent backflow of blood

Four sets of valves allow blood to move through the heart chambers in only one direction. These valves consist of crescent-shaped flaps (or "cusps") of fibrous tissue that are connected to the internal walls of the heart.

The valves are located 1) at the entrance to the pulmonary artery (the pulmonary valves); 2) at the entrance to the aorta (the aortic valves); 3) between the right atrium and the right ventricle (the tricuspid / right atrioventricular valve); and 4) between the left atrium and the left ventricle (the bicuspid / mitral / left atrioventricular valve).

Opening and closing of these valves occurs in response to changes in blood pressure on either side as blood flows through them – so the flaps of the valves separate when blood is flowing in the correct direction, and then close into a tight seal to prevent any backflow.

Structure of the Heart – Schematic

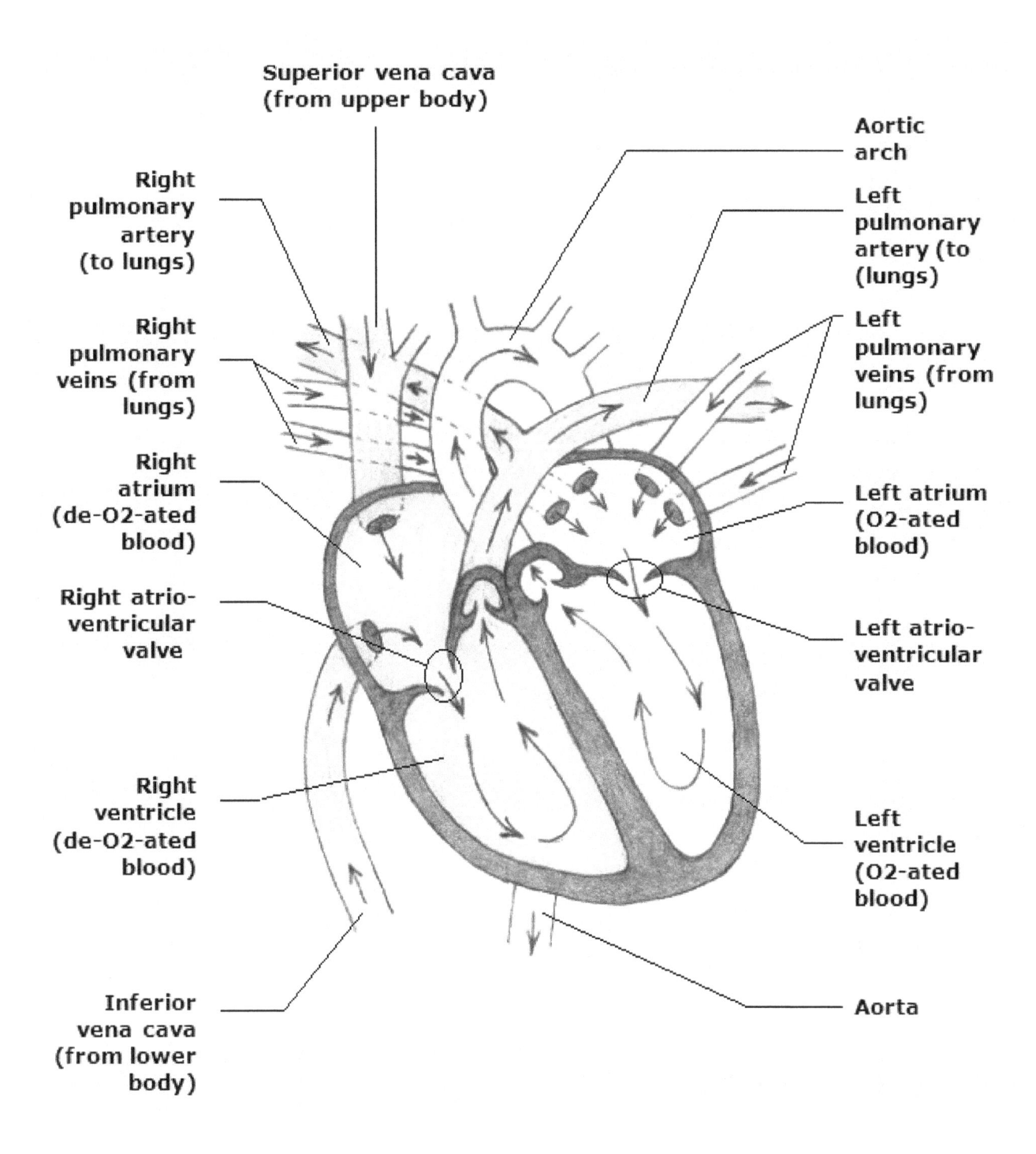

- Superior vena cava (from upper body)
- Aortic arch
- Right pulmonary artery (to lungs)
- Left pulmonary artery (to (lungs)
- Right pulmonary veins (from lungs)
- Left pulmonary veins (from lungs)
- Right atrium (de-O2-ated blood)
- Left atrium (O2-ated blood)
- Right atrio-ventricular valve
- Left atrio-ventricular valve
- Right ventricle (de-O2-ated blood)
- Left ventricle (O2-ated blood)
- Inferior vena cava (from lower body)
- Aorta

O2-ated blood = oxygenated blood
de-O2-ated blood = deoxygenated blood

Structure and Function of the Heart – II

Pulmonary veins
- Deliver oxygenated blood from the lungs into the left atrium of the heart.

Left atrium
- Receives oxygenated blood from the pulmonary veins.

Left ventricle
- Receives oxygenated blood from the left atrium;
- When the heart muscle contracts, oxygenated blood is pumped out of the left ventricle up into the aorta.

Aorta (large systemic artery leaving the heart)
- Carries oxygenated blood around the body to supply all body cells with oxygen.

Venae cavae (large veins entering the right atrium)
- Deposit deoxygenated blood from the body cells into the right atrium of the heart.

Right atrium
- Receives deoxygenated blood from the venae cavae.

Right ventricle
- Receives deoxygenated blood from the right atrium;
- When the heart muscle contracts, deoxygenated blood is pumped out of the right ventricle up into the pulmonary arteries.

Pulmonary arteries
- Deliver deoxygenated blood from the right ventricle of the heart to the lungs;
- Carbon dioxide is transferred from the bloodstream to the lungs to be breathed out;
[- Oxygen breathed in is transferred from the lungs to the bloodstream, to be transported back to the heart via the pulmonary veins.]

The Cardiovascular System – Schematic

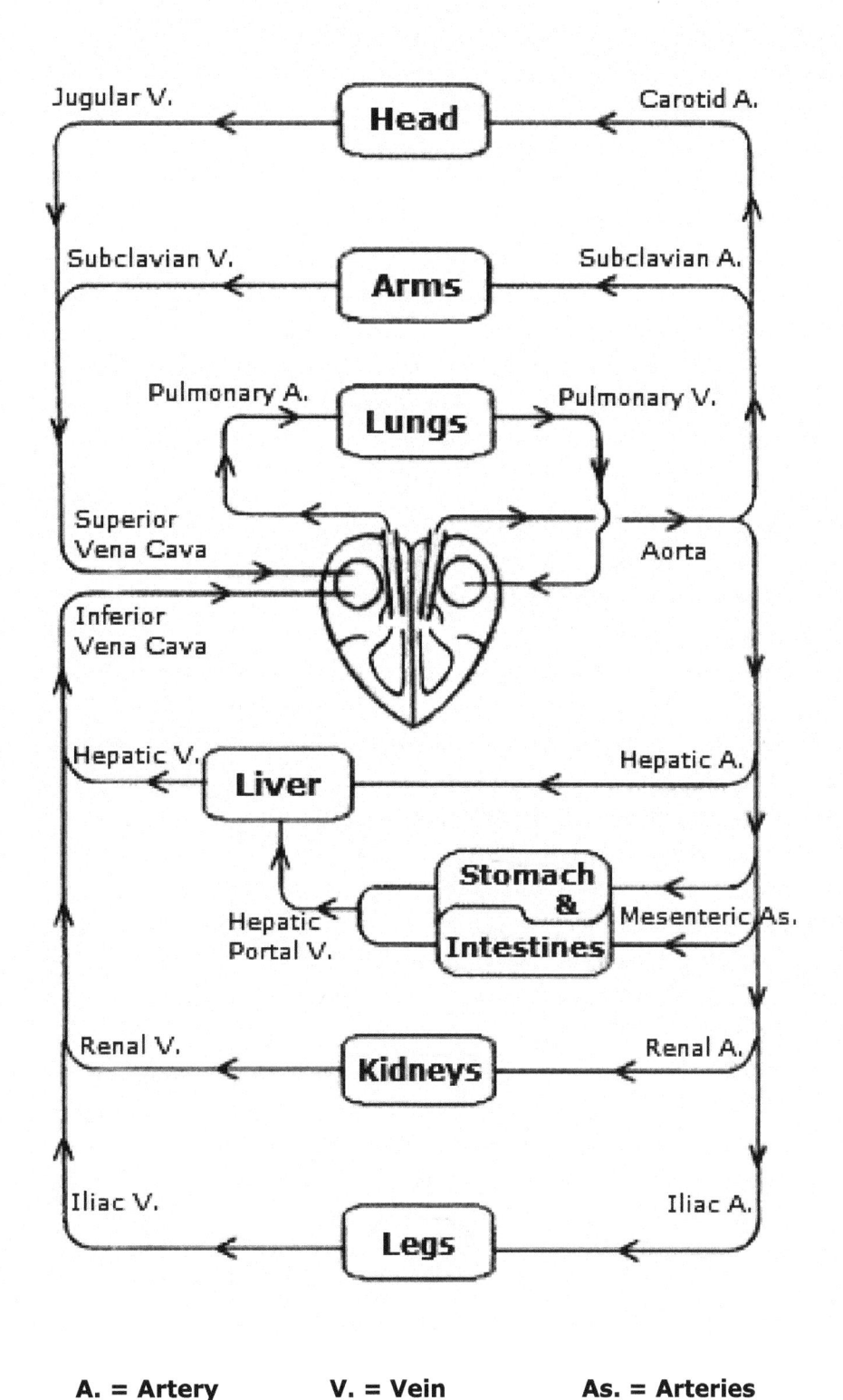

```
Jugular V.  ←——  Head  ←——  Carotid A.
Subclavian V.  ←——  Arms  ←——  Subclavian A.
          Pulmonary A.  →  Lungs  →  Pulmonary V.
Superior Vena Cava                                   Aorta
Inferior Vena Cava
Hepatic V.  ←——  Liver  ←——  Hepatic A.
          Hepatic Portal V.  ←——  Stomach & Intestines  ←——  Mesenteric As.
Renal V.  ←——  Kidneys  ←——  Renal A.
Iliac V.  ←——  Legs  ←——  Iliac A.
```

A. = Artery   **V. = Vein**   **As. = Arteries**

Heart Contraction (Pumping Action) – I

Automatic rhythmic contraction and relaxation
The autonomous (involuntary) contractile properties of cardiac muscle, which makes up the walls of the heart, enable the heart to relax and contract rhythmically in order to first fill with blood, then pump it out around the body (oxygenated blood) or to the lungs (deoxygenated blood).

The rate of relaxation and contraction (the heart rate) is set by two specialised groups of cells located in the walls of the heart. The primary group of cells (the heart's natural "pacemaker"), located in the the upper wall of the right atrium, is called the sinoatrial (SA) node. Electrical signals emanating from here cause the atria to contract. The secondary group of cells, located in the lower wall (floor) of the right atrium, is called the atrioventricular (AV) node. Moments after the signals from the SA node are sent, the AV node sends out signals that cause the ventricles to contract. These specialised groups of cells thus work together to maintain the rhythmic contraction and relaxation of the heart.

The cardiac (or heartbeat) cycle
[Note: Diastole = relaxation/filling. Systole = contraction/emptying.]
1) Diastole / relaxation of both the atria and the ventricles
– Relaxation and filling phase
Deoxygenated blood (from the body cells/systemic circulation) enters the right atrium, and oxygenated blood (from the lungs/pulmonary circulation) enters the left atrium. The blood from each of the atria then flows through into the respective ventricle. By the end of this phase, the ventricles are filled to about 80 percent capacity.
2) Atrial systole (contraction) / ventricular diastole (relaxation)
– Filling of the ventricles to capacity
The atria contract, squeezing any remaining blood from the atria into the ventricles.
3) Atrial diastole (relaxation) / ventricular systole (contraction)
– Pumping of blood from the ventricles into the arteries
The ventricles contract, forcing blood through the pulmonary and aortic valves (between the ventricles and the arteries) into:
a) the pulmonary arteries (carrying deoxygenated blood from the body cells towards the lungs – pulmonary circulation) (blood squeezed up from the right ventricle); and
b) the aorta (carrying oxygenated blood from the lungs towards the rest of the body – systemic circulation) (blood squeezed up from the left ventricle). As this phase ends, diastole starts again.

The illustrations on the following page depict the events occurring in phases 2) and 3) of the cardiac cycle.

Heart Contraction (Pumping Action) – II

The ventricles fill with blood during ventricular diastole (relaxation), and empty during ventricular systole (contraction).

Filling of the ventricles

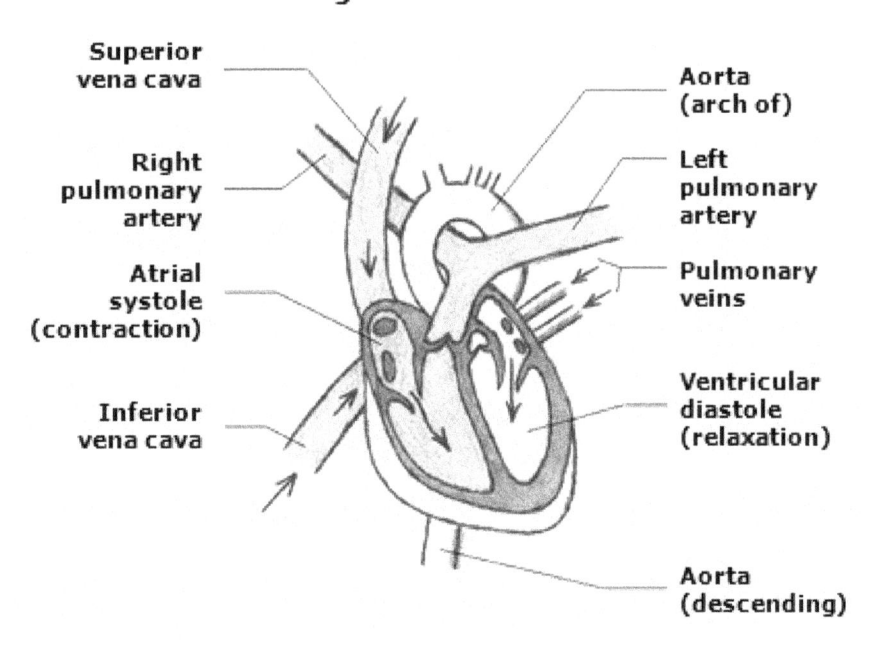

- Superior vena cava
- Right pulmonary artery
- Atrial systole (contraction)
- Inferior vena cava
- Aorta (arch of)
- Left pulmonary artery
- Pulmonary veins
- Ventricular diastole (relaxation)
- Aorta (descending)

Emptying of the ventricles

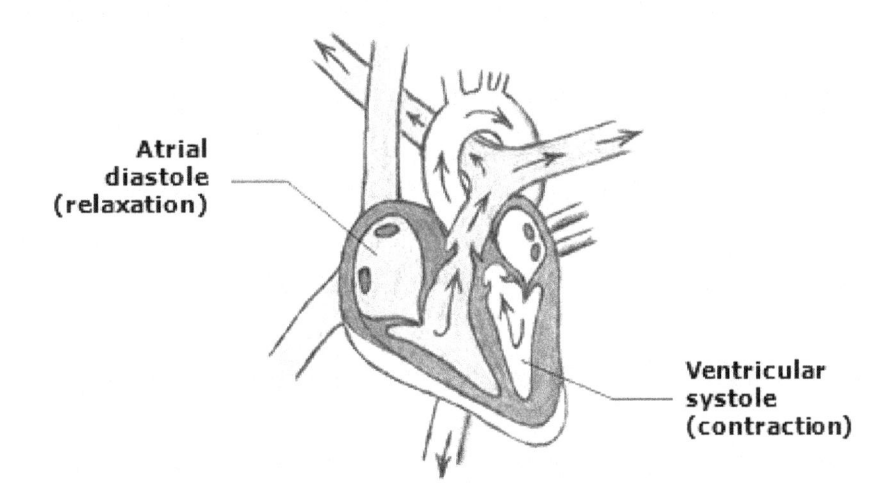

- Atrial diastole (relaxation)
- Ventricular systole (contraction)

The Arterial System – Main Arteries

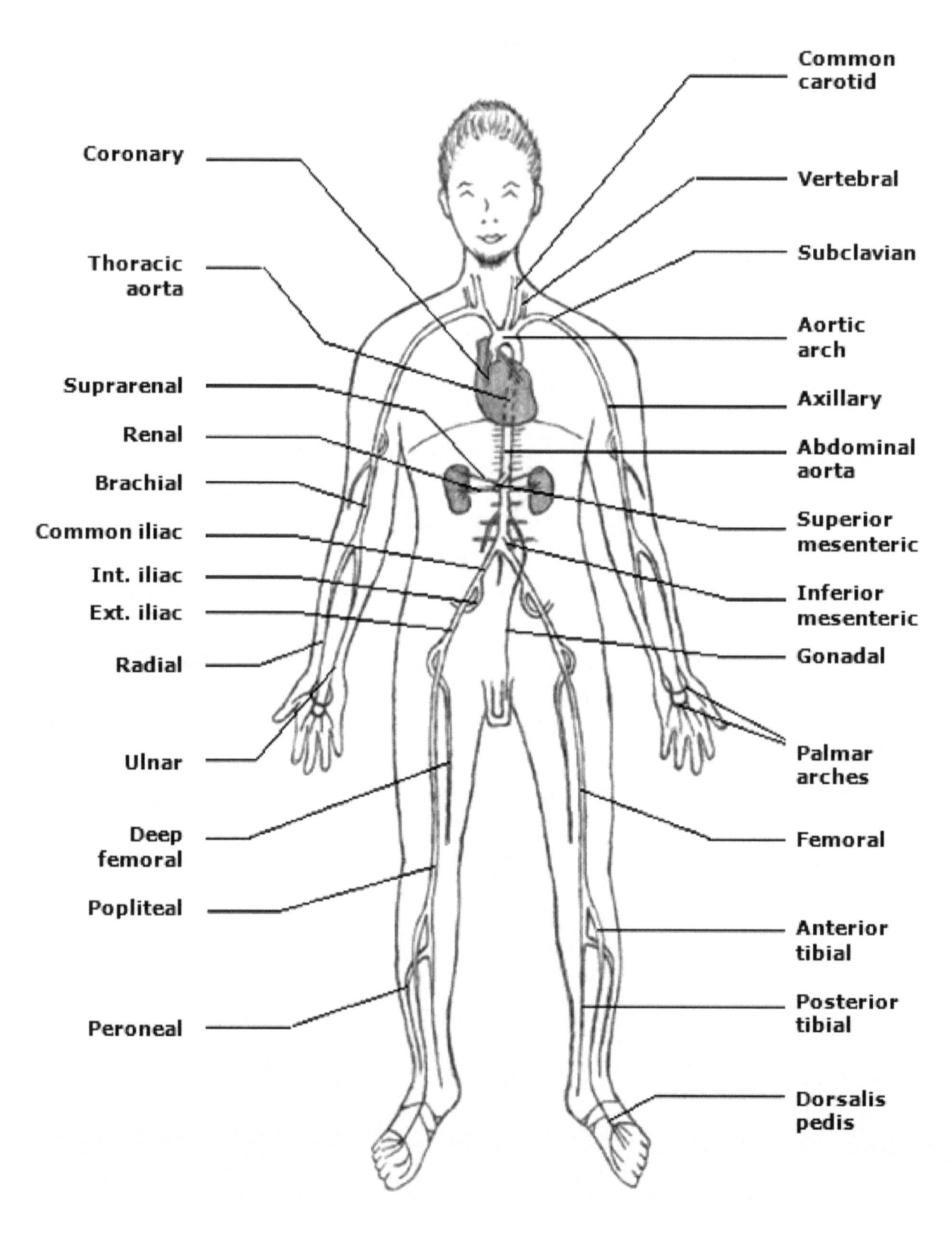

- Coronary
- Thoracic aorta
- Suprarenal
- Renal
- Brachial
- Common iliac
- Int. iliac
- Ext. iliac
- Radial
- Ulnar
- Deep femoral
- Popliteal
- Peroneal

- Common carotid
- Vertebral
- Subclavian
- Aortic arch
- Axillary
- Abdominal aorta
- Superior mesenteric
- Inferior mesenteric
- Gonadal
- Palmar arches
- Femoral
- Anterior tibial
- Posterior tibial
- Dorsalis pedis

The Venous System – Main Veins

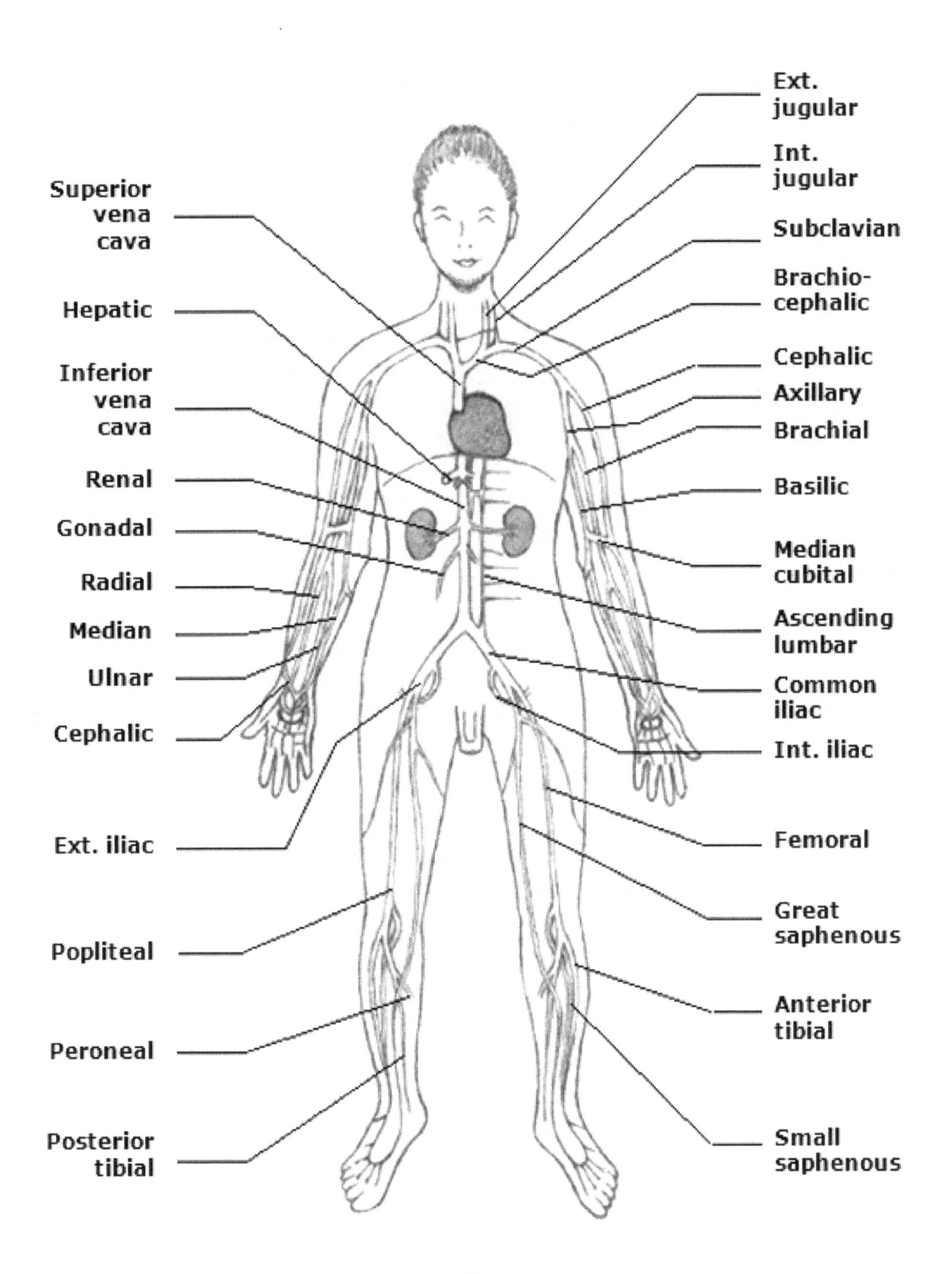

- Superior vena cava
- Hepatic
- Inferior vena cava
- Renal
- Gonadal
- Radial
- Median
- Ulnar
- Cephalic
- Ext. iliac
- Popliteal
- Peroneal
- Posterior tibial
- Ext. jugular
- Int. jugular
- Subclavian
- Brachio-cephalic
- Cephalic
- Axillary
- Brachial
- Basilic
- Median cubital
- Ascending lumbar
- Common iliac
- Int. iliac
- Femoral
- Great saphenous
- Anterior tibial
- Small saphenous

The Cardiovascular System – Part II – Summary

Structure and Function of the Heart

Four heart chambers. The heart is made up of four chambers: the upper chambers are the right and left atria; the lower chambers are the right and left ventricles.

Left side – oxygenated blood. The left side of the heart deals with oxygenated blood received from the lungs, via the pulmonary veins, into the left atrium, and pumps this out through the aorta (from the left ventricle) to supply all areas of the body.

Right side – deoxygenated blood. The right side of the heart deals with deoxygenated blood from all areas of the body, received via the superior and inferior venae cavae into the right atrium, and pumps this out through the pulmonary arteries (from the right ventricle) to the lungs.

Four sets of valves. Four sets of valves allow blood to move through the heart chambers in only one direction. These valves consist of crescent-shaped flaps (or "cusps") of fibrous tissue that are connected to the internal walls of the heart. These valves are located 1) at the entrance to the pulmonary artery (the pulmonary valves); 2) at the entrance to the aorta (the aortic valves); 3) between the right atrium and the right ventricle (the tricuspid valve); and 4) between the left atrium and the left ventricle (the bicuspid, or mitral, valve). Opening and closing of these valves occurs in response to changes in blood pressure on either side as blood flows through them – so the flaps of the valves separate when blood is flowing in the correct direction, and then close into a tight seal to prevent any backflow.

Structures of the Heart

Pulmonary veins: Deliver oxygenated blood from the lungs into the left atrium of the heart.
Left atrium: Receives oxygenated blood from the pulmonary veins.
Left ventricle: Receives oxygenated blood from the left atrium. When the heart muscle contracts, oxygenated blood is pumped out of the left ventricle up into the aorta.
Aorta (large systemic artery leaving the heart): Carries oxygenated blood around the body to supply all body cells with oxygen.
Venae cavae (large veins entering the right atrium): Deposit deoxygenated blood from the body cells into the right atrium of the heart.
Right atrium: Receives deoxygenated blood from the venae cavae.
Right ventricle: Receives deoxygenated blood from the right atrium. When the heart muscle contracts, deoxygenated blood is pumped out of the right ventricle up into the pulmonary arteries.
Pulmonary arteries: Deliver deoxygenated blood from the right ventricle of the heart to the lungs. Carbon dioxide is transferred from the bloodstream to the lungs to be breathed out. [Oxygen breathed in is transferred from the lungs to the bloodstream, to be transported back to the heart via the pulmonary veins.]

Heart Contraction (Pumping Action)

Automatic contraction and relaxation
- The autonomous contractile properties of cardiac muscle (which makes up the heart walls), enable the heart to relax and contract rhythmically (forming the heart rate, or pulse). The heart first fills with blood, then pumps it out around the body (oxygenated blood) or to the lungs (deoxygenated blood).
- The heart rate is set by two specialised groups of cells in the heart walls: 1) the sinoatrial (SA) node and 2) the atrioventricular (AV) node. Electrical signals emanating from the SA node cause the atria to contract; moments later, the AV node sends out signals that cause the ventricles to contract. The SA and the AV nodes work together to maintain the rhythmic contraction and relaxation of the heart.

Heart Contraction (Pumping Action)
(continued)

The cardiac (heartbeat) cycle
1) Diastole (relaxation) of both atria and ventricles. Deoxygenated blood enters the right atrium, and oxygenated blood enters the left atrium. The blood from each of the atria then flows through into the respective ventricle. By the end of this phase, the ventricles are filled to about 80 percent capacity.
2) Atrial systole (contraction) / ventricular diastole (relaxation). The atria contract, squeezing any remaining blood from the atria into the ventricles (filled now to capacity).
3) Atrial diastole (relaxation) / ventricular systole (contraction). The ventricles contract, forcing the blood through the valves between the ventricles and the major arteries into: a) the pulmonary arteries (deoxygenated blood); and b) the aorta (oxygenated blood). As this phase ends, diastole starts again.

The Cardiovascular System – Overview
(Two Circulatory Systems)

The heart effectively links two circulatory systems:
1) The pulmonary circulation: That involving the heart and the lungs (where oxygenated blood travels from the lungs to the heart, and deoxygenated blood travels back from the heart to the lungs); and
2) The systemic circulation: That involving the heart and the rest of the body (where oxygenated blood travels from the heart to all areas of the body, and deoxygenated blood returns from the body cells back to the heart).

The Arterial System – Main Arteries

Neck: Common carotid, vertebral. **Trunk:** Aorta (thoracic and abdominal sections), pulmonary, subclavian, suprarenal, renal, superior and inferior mesenteric, common, internal and external iliac, gonadal. (**Heart:** coronary.) **Arm:** Axillary, brachial, radial, ulnar. (**Hand:** palmar arches.) **Leg:** Femoral, deep femoral, popliteal, anterior and posterior tibial, peroneal. (**Foot:** Dorsalis pedis.)

The Venous System – Main Veins

Neck: External and internal jugular. **Trunk:** Superior and inferior venae cavae [Note: plural of vena cava], pulmonary, brachiocephalic, subclavian, hepatic, (hepatic portal), renal, gonadal, ascending lumbar, common, internal and external iliac. **Arm:** Cephalic, axillary, brachial, basilic, median cubital, radial, median, ulnar. **Leg:** Femoral, great and small saphenous, popliteal, anterior and posterior tibial, peroneal.

* * *

Quiz – The Cardiovascular System – Part II

1. The upper chambers of the heart are called the right and left ……………………………………

2. The lower chambers of the heart are called the right and left ……………………………………

3. The left side of the heart deals with ………………………… blood, whereas the right side of the heart deals with ……………………………… blood.

4. The flow of blood through and out of the heart is controlled by ……………….. [number] sets of valves.

5. The valve between the right atrium and right ventricle is called the …………………. valve.

6. The valve between the left atrium and left ventricle is called the …………………… valve.

7. The large veins that pour deoxygenated blood into the right atrium are called the ……………………………………..

8. The veins that pour oxygenated blood into the left atrium are called the ………………………………………

9. The artery exiting from the left ventricle to carry oxygenated blood to the body cells is called the …………………………

10. The arteries exiting from the right ventricle to carry deoxygenated blood to the lungs are called the ………………………

11. The specialised groups of cells in the walls of the heart that maintain the automatic, rhythmic contraction and relaxation of the heart (the heart rate) are called the ………………….. and the ……………………...

12. Diastole means ………………………………..

13. Systole means ………………………………...

14. The ventricles fill to capacity during ventricular ………………………….. and empty during ventricular ……………………………..

* * *

Answers to Quiz Questions – Module 2

Section 1 – The Muscular System – Part I
1. skeletal, smooth, cardiac. **2.** skeletal, smooth, cardiac. **3.** muscle fibre. **4.** muscle fibres, myofibrils, myofilaments. **5.** actin, myosin. **6.** tendons. **7.** contract. **8.** contractile, actin, myosin.

Section 2 – The Muscular System – Part II
1. flexion. **2.** abduction. **3.** pectoralis major muscles (or pectorals). **4.** deltoid. **5.** proximal. **6.** triceps. **7.** 4, quadriceps. **8.** trapezius. **9.** gluteus maximus. **10.** hamstrings. **11.** gastrocnemius. **12.** Achilles.

Section 3 – The Cardiovascular System – Part I
1. transport of essential substances, transport of waste materials, helps to fight infection, prevents loss of body fluid after injury (blood clotting function), regulation of body temperature. **2.** white blood cells (or leukocytes), platelets (or thrombocytes). **3.** oxygen. **4.** white blood cells (or leukocytes). **5.** platelets (or thrombocytes). **6.** away from, towards. **7.** oxygen. **8.** carbon dioxide. **9.** capillaries. **10.** systemic. **11.** pulmonary.

Section 4 – The Cardiovascular System – Part II
1. atria. **2.** ventricles. **3.** oxygenated, deoxygenated. **4.** 4. **5.** tricuspid. **6.** bicuspid (or mitral). **7.** venae cavae. **8.** pulmonary veins. **9.** aorta. **10.** pulmonary arteries. **11.** sinoatrial node, atrioventricular node. **12.** relaxation. **13.** contraction. **14.** diastole, systole.

About the Author

Amanda Jackson-Russell originally trained as a medical scientist, qualifying with a B.Sc. (Hons) in Physiology from the University of Manchester, England, and a Ph.D. in Neuroscience from the University of Southern California, Los Angeles. She worked in medical research, then medical and health publishing, and subsequently maintained a successful freelance writing and editing career for over 20 years. In parallel with her scientific and publishing work, Amanda also followed her interests in natural and complementary medicine, gaining a number of practitioner qualifications including in massage therapy, Reiki, spiritual healing, aromatherapy, emotional freedom techniques (EFT) and hypnotherapy. She has also been a qualified yoga teacher with the British Wheel of Yoga for over 25 years. Amanda currently practices as a clinical and cognitive-behavioural hypnotherapist and stress management consultant, and continues to carry out freelance writing and editing part-time.

During her long and varied career, Amanda became aware of a number of issues many students, therapists and teachers of paramedical subjects (including yoga teachers) face, particularly when needing to learn essentials of anatomy and physiology for their professional training, and then maintain knowledge of the key elements for their practices. There appeared to be few helpful publications available that provided accurate and easy-to-follow information without the reader being burdened with or getting "bogged down" in unnecessary depth and detail of information. In this book series *Essentials of Anatomy & Physiology A Review Guide for Therapists, Yoga Teachers, Holistic Health Practitioners, Healers & Wellbeing Coaches,* Amanda has endeavoured to present review guides covering all the main areas of anatomy and physiology that should be helpful to qualified and student therapists, wellbeing coaches, yoga teachers and others interested in complementary, natural and holistic health.

Other Modules in this Series

Essentials of Anatomy & Physiology – A Review Guide for Therapists, Yoga Teachers, Holistic Health Practitioners, Healers & Wellbeing Coaches
by Amanda Jackson-Russell, Ph.D.

Module 1
Cells & Tissues
The Skin
The Skeletal System, Parts I, II & III
Published: November 2019

Module 3
The Nervous System, Parts I, II & III
The Respiratory System
Publication due: July 2020

Module 4
The Digestive & Eliminative System
The Lymphatic System
The Endocrine System
Publication due: August 2020

Module 5
The Immune System
The Reproductive System
The Urinary System
Publication due: October 2020

Lightning Source UK Ltd.
Milton Keynes UK
UKHW021115221022
410882UK00009B/1158